PRESENT AT THE
FUTURE

PRESENT AT THE

FROM EVOLUTION
TO NANOTECHNOLOGY,
CANDID AND CONTROVERSIAL
CONVERSATIONS
ON SCIENCE AND NATURE

FUTURE

IRA FLATOW

COLLINS
An Imprint of HarperCollins Publishers

For Mike Waters

HarperCollins books may be purchased for educational, business, or sales
promotional use. For information, please write: Special Markets Department, HarperCollins Publishers, 10 East 53rd Street, New York, NY 10022.

First Collins paperback edition published 2008.

Designed by Mary Austin Speaker

The Library of Congress has catalogued the hardcover edition as follows:

Flatow, Ira.
    Present at the future : from evolution to nanotechnology, candid and
controversial conversations on science and nature / Ira Flatow.—1st ed.
        p. cm.
    Includes index.
    1. Science—Miscellanea.   I. Title.
    ISBN 978-0-06-073264-6

    Q173.F59 2007
    500—dc22
                                                            2007014583

    ISBN 978-0-06-073265-3 (pbk.)

08 09 10 11 12 ID/RRD 10 9 8 7 6 5 4 3 2 1

CONTENTS

✳

INTRODUCTION

✳

One of the joys of being a science journalist is the "aha!" moment. That brilliant flash of light, that moment of epiphany, when all the cylinders in your head click and you come to understand something you had never understood before. Sometimes your mouth may actually fall open as your eyes wander around the room, not really seeing anything but not under conscious control of your brain, which has just latched on to an idea that had eluded it for quite some time.

As I say, one of the joys of being a science journalist, especially the host of a popular radio program such as *Science Friday*, is the ability to share that moment with millions of others. And one of those moments, among many others, stands out in my mind. It involves solving a problem I was having with one the most widely accepted "truths" about the world around us: why airplanes fly.

Like every science journalist who ponders how things work, I was more than casually acquainted with the commonly accepted ideas about the physics of flight. After all, the explanation for why airplanes fly is one of those unquestioned beliefs in science education, a dogma that has been repeated for generations and taught in eighth-grade science class along with why the sky is blue. (Oh, you were out that day?) In fact I have mouthed the simple explanation countless times on radio and television talk shows. I had sung its praises in my own (previous) books. High-school teachers who coach award-winning teams in science competitions would all say the same thing: "The basis for why airplanes fly is a concept in physics called Bernoulli's principle. We all know that." No-brainer. Period. End of discussion.

Just what is Bernoulli's principle? It's quite simple to state: Daniel Bernoulli, a Swiss mathematician, discovered in 1733 that the faster a fluid flows, the lower its pressure.

And that explanation has been used ever since the 1930s to explain why an airplane flies. In short, the explanation states that because of the design of a wing, air will fly faster over the top than the bottom. According to Bernoulli, faster-flowing air on top means less pressure on top. More pressure on the bottom than on the top lifts the wing. So the wing is literally sucked up into the sky, the same way you suck up soda in a straw. Complex science-fair projects and even simple strips of paper over which you blow—the paper flies up—have been used for decades to demonstrate Bernoulli's principle. And as I say, I too, when asked to explain in simple terms what makes an airplane lift off the ground, would say the same thing: Bernoulli's principle.

That all came to an end one day in 1989. After that day, I would no longer freely mention Bernoulli and flight in the same breath, unless it was to denounce the principle. Because on that day I met a man who would change my life forever. And I vowed to set the record straight.

I had just finished making a speech to a group of folks at a science conference at the University of Vermont. After polite applause and a brief question-and-answer period, people began to file out. Just as I was about to collect my papers and head for the door, the real answer to why airplanes fly appeared before me in the form of Norman Smith. Norman Smith had spent more than two decades as a research aerodynamicist for NASA and before that for its pre-space predecessor, the National Advisory Committee for Aeronautics. He was the author of more than a dozen science books as well. And now in his later years, after a very successful career in aviation, he was on a one-man crusade: to get rid of Bernoulli's principle as the explanation for why airplanes fly.

"I listened to your speech and was very impressed with what you had to say," said Smith. "Except for one thing. That explanation about why airplanes fly."

"Oh. Did I say something wrong?"

"Yes, you did. That explanation you gave, the one using Bernoulli's principle, is no longer acceptable. It's old and not really accurate."

Smith was trying to be polite. What he would have really liked to have said, I think, was "You dummy, get with the program. Here's another guy I'm going to have to reeducate."

Education was what Smith did best. As far back as 1972, in a book and in an article published in that most sacred of all science-teacher magazines, *The Physics Teacher,* Smith met the enemy wherever he found it. Wherever teachers banked on Bernoulli to teach the theory of flight—in newspapers, in magazines, in a lecture hall—Smith was there to remove Bernoulli from its central role in flight. He called it the "Bernoulli myth" and said it was "the most persistent, pernicious error in school science books."

Not only is Bernoulli used to explain aircraft lift, Smith lamented, but "I have found articles on ornithology that borrow the error to explain bird-wing lift!"

Smith handed me a copy of one of his magazine articles from *The Physics Teacher* and asked me to read and see for myself.

"The math is pretty easy for a smart guy like you."

By now I was more than intrigued. *Embarrassed* might be a good word. Here I had just lectured to hundreds of people—I was a role model, an author, an "expert"—and now, if Smith was right, I was an idiot. Smith could see my sheepish look, and his smile told me *Not to worry; you've merely made a mistake common to physics teachers around the world. Join the crowd.*

"Just what *is* the right explanation, then?" I needed to hear his take.

"Very simple: Newton. And his laws of motion. You can easily—and correctly—explain why airplanes fly from first principles. No need to resort to Bernoulli. He was created—really pulled out of a hat—around World War II when the airplane was becoming popular and people wanted a simple explanation. But in reality it takes more time to explain the complicated workings of Bernoulli's principle than it does the simple laws of Newton. In this case it's very simple: Airplanes fly because the wing makes the air go down, so the airplane goes up. Action—reaction. Newton's third law. How hard is that to understand?"

I took Smith's papers home and read them. They made a lot of sense. I researched Smith and uncovered other articles about him and his quixotic quest to set the record straight. Jerry Bishop, a highly respected colleague of mine and science writing icon of the *Wall Street Journal*, had come across Smith in 1972 and had written a major *Wall Street Journal* article about the quest. Being the great journalist that he is, Jerry presented the case but never reached a conclusion. He let the reader decide.

But I was convinced. Of course it was Newton! The beauty of flight was in this detail.

Thus began my own crusade to replace Bernoulli with Newton, just as Smith had tried to do. I soon found and talked with other like-minded science teachers who had begun to make their own contributions. They began to question the very basics of accepted textbook ideas about how air flows over a wing, and they discovered that they were wrong.

I chatted with airplane pilots who had never once relied on Bernoulli to fly. They talked about "angle of attack" and "stalling the wing" and "airspeed" and such. A helicopter pilot showed me his "movable wing" plane—his copter—and told me to take a close look at its shape. He pointed to the end of the wing and said, "Does this look like your textbook Bernoulli airfoil? Of course not: it's symmetrical. Contrary to Bernoulli dogma, wings don't have to be rounded on top and flat on the bottom. See? Mine are symmetrical, rounded on top as well as bottom. And I'm getting off the ground!"

I decided to take flying lessons to learn the details firsthand. My flight instructor told me that the Federal Aviation Administration (FAA) still requires using Bernoulli's principle as a teaching aid, despite the growing belief that it's an inadequate explanation. It's even on the FAA exam. My flight textbooks and CDs all had Bernoulli. But not once during my training, not on takeoffs or landings, not when recovering from a stall or learning how to "trim for level flight," did the word *Bernoulli* come up.

So I knew that in writing this book I had the opportunity to set the record straight. No longer would Bernoulli be the centerpiece of any chapter about why airplanes fly. And for me, a baseball fanatic, there was an even greater injustice: the use of Bernoulli's principle to explain why a curveball curves or a fastball "rises." I too bow my head and plead "guilty as charged." But no longer.

So if the section in this book titled "Why an Airplane Flies: Debunking the Myth" appears more detailed or perhaps more passionate than the others, please bear with me. I'm getting a lot off my chest. I've spent almost 15 years thinking about this, and we all know there is no greater crusader than a reformed sinner.

This episode serves as a case in point about another myth: that science knows everything. I'm reminded of the letter I received the first week that *Science Friday* was on the air in 1991. A woman—let's call her Barbara from New Jersey—wrote that she had just listened to a program we had broadcast about the extinction of the dinosaurs and the possibility that they had been wiped out by a comet or asteroid hitting the Earth. On that show, a scientist with a competing theory phoned in to say that he disagreed with this new asteroid theory about extinction. The caller got into, shall we say, a frank exchange of views with my guest about their competing theories. They got into a professional "disagreement." Barbara wrote that she was shocked to hear scientists arguing. She had never heard researchers disagree. "Isn't that what science is for? Science *knows* the truth, doesn't it?" Imagine that! Scientists arguing! A concept, I think, foreign to most laypeople.

After watching science do its thing for a while, you realize that knowledge is really a moving target. What we know today will probably be wrong tomorrow. And science is that tool for discovery. When science tells us something, chances are that it will tell us something different a few years from now.

And that's what makes truth stranger than fiction.

PART I

—

THREE POUNDS
OF GRAY MATTER

CHAPTER ONE

THE MIND'S WINDOW

A Fox enters the storeroom of a theater. Rummaging through the contents, he is frightened by a face glaring down on him. But looking at it closely he discovers it is only a mask, of the kind worn by actors. "You look very fine," says the Fox. "It is a pity you haven't any brains."

— AESOP'S FABLES

Is it possible to understand our minds? To understand what consciousness is all about? What happens in our brains when we learn or remember? What goes on when we "enjoy" or feel "depressed"?

Understanding the complex biochemistry that turns electrical and chemical energy into thoughts, memories, and feelings is one of the greatest challenges of science. Neuroscience has become "the neurosciences" as genetics, physics and engineering, pharmaceuticals, psychology and psychiatry, and computer science have gotten into the act and contributed to what we know. But the brain is so complex that neuroscientists have a long way to go before we can understand completely how the brain works.

We know that people have been fascinated with the brain for at least 6,000 years. About 4000 BC, an anonymous writer put down the very first observations of how the brain works, anticipating *The Wonderful Wizard of Oz* by noting that eating poppies induced feelings of euphoria and well-being. It seems that people also have always been hitting their heads: The ancient Egyptians documented on papyrus medical treatments for 26 different kinds of brain injury, and pre-Inca civilizations practiced primitive brain surgery, probably for mental illnesses, headaches, or epilepsy. In the Middle Ages, many people witnessed miracles, wonders, and visions—perhaps because they didn't realize that their brains were tripping on LSD. In 1938, Dr. Albert Hofmann, a distinguished Swiss chemist who was interested in the medicinal properties of plants, was studying the ergot fungus at Sandoz Pharmaceuticals in Basel. He found that ergot contained a kind of lysergic acid with hallucinatory properties, an acid that Hofmann synthesized into LSD for the first time. Ergot fungus often affects grain. In medieval times, grain with the fungus could have been milled into rye bread, causing hallucinations—and along with them, superstitions and religious fervor that could have been due to altered brain chemistry.

Today, we know that the brain runs on electricity—though not the kind of electricity that lights up the lamp over my desk or runs my computer. I'm talking about bioelectricity, which allows the neurons, or cells in the brain, to communicate with one another. Every living cell functions with electricity. When food is digested and turned into blood sugar, or glucose, and dissolves in water inside a cell, its atoms lose or gain electrons. They become free-floating particles called ions, which have either a positive or a negative charge. Since electricity is charge in motion, the movement of charged ions inside a living cell is electricity. When ions move, there is a corresponding shift in charge, an electrochemical change that produces an electric charge, the nerve signal. Every fraction of a second, each nerve cell in the brain and body receives signals that prompt it to respond or not. When a neuron sends a message to another neuron, the signal moves along as a traveling electric pulse. Recently I saw a photo of a neuron hooked up to a nanoscale plastic circuit on a chip—just one experiment in nanotechnologists' efforts to build a super-tiny transistor no bigger than a molecule.

While ancient peoples thought that epilepsy was caused by demonic possession, we know that an epileptic seizure is an outward sign of abnormal electrical activity in the brain, due to an imbalance in neural activity that leads to an increase in the rate of neural firing, which can then spread to other parts of the brain. But there are still so many mysteries left, especially how our memories, our hopes and dreams, our intelligence, and everything else we're thinking of when we say *mind* are encoded in our brains.

MAKING CONNECTIONS

One of the biggest mysteries about the brain is how it begins. When a fetus is only one month old, its first brain cells, or neurons, are growing at the mind-boggling rate of 250,000 neurons a minute. Eventually, those neurons form literally trillions of connections, called synapses, between cells. These connections are well organized, not random: Each neuron finds its correct place in the brain. By the time a baby is born, it has 100 billion neurons, and its brain looks very much like an adult's. It's more developed than any other part of the baby's body, and it's disproportionately large. After birth, the brain begins to be shaped by environment—the world around the infant and the baby's experiences. Newborns spend more than 20 percent of their sleep in rapid eye movement (REM) sleep, which some researchers think involves a kind of learning process. Neurologists are studying how the brain shapes itself in response to the demands the environment makes on it. They know the brain changes over a person's lifetime, as it thinks, controls muscles and limbs, learns, and remembers. The billions of neurons in a person's brain continually connect and reconnect on many different levels, in response to what their owner does and experiences.

Some of the things we don't know about the brain are surprisingly basic. One thing that babies and very young children do a lot is sleep. In fact, they spend half their childhood asleep—and every parent knows how important that is and what their kids can be like if they don't get their naps. Adults spend about a third of their time asleep, and that doesn't appear to be an enormous waste of valuable time. Experiments where people have tried to stay awake for as long as 200 hours have induced hallucinations and paranoia. If adults have troubling sleeping—and according to a 2005 poll from the National Sleep Foundation, 57 percent of Americans do—nearly every aspect of their lives is affected, leaving them prone to making mistakes at work, having car accidents, being too sleepy for sex.

Sleep, obviously, offers muscles and other parts of the body a chance to rest. But not the brain: measurements of its electrical activity reveal that it's hardly dormant. Still, until 1953, when two researchers who were studying children's sleep patterns described REM sleep for the first time, scientists assumed that the brain was inactive during sleep. But exactly why sleep is so important, regardless of your age, and what sleep means to the brain are questions scientists cannot answer yet. Perhaps sleep helps consolidate learning (more on that later in "Sleep and Learning: Caffeine in Your Beer," on page 43).

TEENAGERS: WHAT WERE YOU THINKING!

In a child's first year, the brain triples in size, until it's almost three quarters of the size of an adult's. The brain achieves its full growth at about age 17. The number of neurons doesn't increase, but the number of synapses do as children imitate, learn, remember, and add to their experiences. By adulthood, the brain has 100 trillion synapses. But before adulthood comes adolescence, when the brain is flooded with hormones. Neurologists only recently have confirmed what every parent knows: The teenage brain is indeed different. In the teenage brain, the prefrontal cortex—the center of reasoning and impulse control—is still forming. In some people, that maturation may not occur until they are 25 years old. That's why so many teenagers have trouble understanding the consequences of their impulsive, destructive behavior. That's why, as every parent can tell you, teenagers are impulsive, emotionally erratic, and liable to make poor decisions. The answer is simple: The section of the brain that can foresee the future, the part that can predict the consequences of actions, is not fully developed.

Every parent also knows that teenagers have a terrible time getting up in time for school and on weekends, preferring to sleep into the afternoon. Neuroscientists now know that teens aren't being lazy—they really do need more sleep. Some schools have even pushed the start of the school day later to take into account the adolescent need for sleep.

SEEING INSIDE THE BRAIN

We've been able to learn more about the adolescent brain because today, we don't have to drill into the brain to find out about it, as early peoples did. We have new technology that allows us to see what the brain looks like and what it's doing. And that's an excellent advance because the brain is delicate, and once a neuron is destroyed, it's gone forever—though the brain often is able to compensate for the loss. Neurosurgeons even have a pithy saying: "You're never the same once air hits your brain."

In 1917, László Benedek, a Hungarian neurologist, took the first pictures of the living brain with what was then the brand-new technology of X-rays. German physicist Wilhelm Roentgen won the Nobel Prize for discovering X-rays, in 1895, making it possible to see structures inside the body without surgery. By 1936, Benedek had come up with the idea of making three-dimensional X-ray pictures of the brain. But X-rays can only show large features of the brain, such as tumors; they cannot show the different layers inside the brain. Nor can they capture the brain's activity and changes, which last only a split second.

In the 1970s, Benedek's idea had been developed into the computerized axial tomography (CAT) scanner—the CAT scan, which uses X-rays to show details of the brain's soft tissue. (Now it's called a computed tomography, or CT, scan.) With a burst of radiation, a CT scan takes a two-dimensional "slice" of the brain, and by putting two-dimensional X-rays together with the aid of a computer, researchers can obtain a three-dimensional image, as Benedek envisioned. Images of the brain from a number of X-rays are digitized, and then reconstructed so that researchers have a cross-sectional view of any part of the brain. CT scans can detect brain damage and can measure brain activity by monitoring blood flow as the patient performs a task. But just like Benedek's early photos, CT scans have limits beyond revealing the brain's structure.

By the time X-rays were discovered, scientists were aware that living brains produce electrical activity. But not until the late 1920s did an Austrian psychiatrist, Hans Berger, actually record this activity for the first time. Berger made an electroencephalograph, another noninvasive way of studying the brain. Electrodes fastened to a patient's scalp pick up the electrical signals produced by the brain and send them to galvanometers, instruments that detect and measure small electrical currents. Like seismometers, which measure earthquakes, galvanometers used to be hooked up to pens that moved over a roll of graph paper to record the characteristic patterns of the current from the patient's brain. Today, the patterns appear on a computer monitor, but electroencephalograms (EEGs) still allow scientists to monitor split-second brain activity and changes. EEGs can tell physicians whether you're awake, asleep, or anesthetized—because your brain patterns look different in each of these states.

During the 2005 debate over Terri Schiavo, the 41-year-old woman in Florida who had spent 15 years in a post-coma, or vegetative, state after a stroke, some physicians who had examined her reported that "her EEG is flat." That meant that her brain was not producing any electricity at all. Since it was no longer functioning, its structure was deteriorating and filling with fluid so that it resembled the inside of a grapefruit. So EEGs can tell us a lot about the state of the brain, but they can't tell us what regions of the brain do what.

To look deeply into the brain, we now have positron emission tomography (PET) scans, which neuroscientists use frequently. To take a PET scan, a neurologist injects into a patient's bloodstream a tiny amount of a radioactive substance, attached to glucose molecules that brain activity absorbs as fuel. In brain tissue, the glucose molecules give off gamma rays, recorded by sensors and then analyzed by computers to picture just where in the brain more glucose-molecule fuel is being used and where less is required. The result is a color-coded map of the brain, where red or yellow usually shows the more active areas that are using more fuel and blue indicates the less active areas that are consuming less fuel. The PET scan of a patient with Alzheimer's disease, for example, is often mostly blue. Right now, PET scans can measure what is happening at 30-second intervals in a tiny portion of the brain. That, of course, is still not fast enough to keep up with brain activity.

In 1977, there was a major breakthrough in brain imaging, the invention of functional magnetic resonance imaging, or fMRI. When you undergo magnetic resonance imaging (MRI), you lie on your back on a movable bed that slides into a giant circular magnet. MRI, like the other kinds of brain scans, isn't at all painful—just uncomfortably noisy once the machine is turned on and begins generating a strong magnetic field. What happens is that the molecules in your body, including your brain, begin to behave like tiny magnets. The MRI machine's magnetic field realigns the hydrogen atoms in your body so that instead of spinning in different directions, they all spin along the same axis, along the length of your body. Now the protons in the hydrogen atoms are facing either up toward your head or down toward your feet. The hydrogen atoms' opposite directions means that most of them cancel out each other's electrical charge. But a few remain, and when the machine sends a beam of radio waves to your brain or to the part of your brain that's being scanned, the pulse makes them resonate and give off radio signals of their own. When the machine shuts off the pulse, the hydrogen atoms return to their normal alignment and release energy, giving off a signal. The machine's sensors detect these signals and feed them into a computer, which generates an image of the different types of tissue in the brain.

Magnetic resonance images are good at detecting changes in the brain, which occur every time you learn something new. MRI has been used to help education researchers develop new teaching methods that have proven to help elementary-school children— most of them are boys—with dyslexia and other reading difficulties. Magnetic resonance images showed researchers that parts of the brains of struggling readers were different from those of successful readers. In one study at the University of Washington, researchers took magnetic resonance images while struggling readers pronounced certain words while lying in the machine. Then the boys were trained in a special curriculum designed to help them read more easily. At the end of the program, the boys underwent MRI again while they read words aloud. The images showed that their brains now looked more like those of normal readers—proof positive that the innovative reading training that they had been given worked. At Yale University, another research team is taking magnetic resonance images of dyslexic readers throughout their lives, for a long-term study of how the brain changes.

Magnetic resonance images are often very colorful and beautiful, as well as clear and detailed. MRI is faster than PET, but it still can't keep up with the extremely rapid changes inside the brain and give us the best possible picture of the brain at work.

Now a new imaging technology in limited use can record brain activity by the millisecond. Magnetoencephalography, or MEG, is still extremely expensive, so there are only a few of the new machines that the imaging requires in existence. Because your brain—like your entire body—works by electricity, it produces a magnetic field. MEG works by detecting the very faint magnetic fields generated by the tiny electric currents from your neurons that are recorded on EEGs. When you have a MEG scan, you sit under a big, very heavy machine that positions magnetic detection coils bathed in helium over your head. The helium chills the coils to supersensitive, superconducting temperatures. Your brain's magnetic field induces a current in the coils that in turn induces a magnetic field in an instrument called a superconducting quantum interference device, or SQUID. The magnetic field can be translated into computer-processed images that provide the most accurate monitoring and timing of brain-cell activity. Of all imaging technologies, MEG provides the best information, so let's hope that it will become cheaper and more available in the future.

Besides imaging, we're learning a lot about the brain from developments in genetics. The Human Genome Project has linked certain genes with normal brain function, such as learning and memory, and with some mental disorders as well. By adolescence, the effects of genes become apparent—including genetically related dysfunctions such as depression or schizophrenia. We know that mental illnesses like these can begin in adolescence, and we can see the changes they cause in the brain in imaging. Pharmaceutical companies have developed new drugs that change brain chemistry in positive ways and help relieve the suffering of people with depression or schizophrenia. But some doctors and philosophers worry that if we know how to explain the brain, we can heal it but also may manipulate it. When the new types of antidepressants that affect the brain's level of serotonin first became popular in the 1990s, some wondered if people would have drugs but not personalities.

The debate over whether to give teenagers drugs that affect brain chemistry illustrates the pitfalls of treating brains that are still developing. For example, teenagers who take antidepressants may be more prone to violent acts, such as suicide or murder. The Native American teenager who killed 10 people at his high school in Red Lake, Minnesota, in 2005 was taking Prozac. While no direct connection between his ingestion of Prozac and his rampage exists, it does bring up disturbing questions about kids and psychotropic drugs. Very little is known about the effects of adult drugs on teens. But that's another story. . . .

At the Neuropsychiatric Institute at the University of California, Los Angeles, researchers have developed their own advanced version of electroencephalography, called cordance or quantitative electroencephalography (QEEG), to link brain function and medication side effects. In studying teenagers who might be susceptible to antidepressants' side effects, the researchers used QEEG to find changes in brain activity in the prefrontal region. They think that when these changes show up before a teenager begins taking an antidepressant, the patient could be vulnerable to side effects. When we are adults, our brains are no longer so soaked with hormones. That can mean that we may need a little help feeling like teenagers.

THE CUTTING EDGE

The National Science Foundation has made neuroscience research a top priority for the twenty-first century, and there have been plenty of new studies because now researchers can use noninvasive technology such as PET scanning and fMRI to look inside the brains of living people and see how they change. Among the surprises they've found is that your brain looks different depending on geography. If you live in Washington State, for example, your brain scan will look different than that of your cousin who lives in Florida. And if you yourself move from Washington to Florida, your scan will look different.

We've learned that the hippocampus, a part of the brain, plays a crucial role in the formation of new memories about personal experiences and in navigation. If you have Alzheimer's disease, your hippocampus will be one of the first parts of your brain to deteriorate.

That's why patients with Alzheimer's disease usually don't recognize close family members and usually easily lose their way.

But that doesn't have to happen to you. Researchers have found out that neurons in the hippocampus do regenerate, deep into old age. They used to think that people relentlessly shed neurons as they aged. But it turns out that there's a constant cycle of neurogenesis, and that's why the brain is plastic, able to change and adapt and even heal sometimes after a serious injury such as a bullet wound or stroke. Your brain cells go through a constant cycle of growth and replacement that continues from birth to death. Some parts of the brain may take over the function of other damaged parts.

SEX ON THE BRAIN

One more piece of recent news about the brain: "Sex matters." That's what the National Academy of Sciences said in 2001 in a major report on how sex differences affect human health. The academy's adage applies to the workings of the brain as well. There *are* significant differences between male and female brains. We still don't know why men forget their loved ones' birthdays or won't ask for directions on road trips, while women readily remember important dates and ask for directions. But we do know for sure that women's brains are smaller. In January 2005, the president of Harvard University sparked major protests, especially from women scientists and engineers, by publicly speculating that differences in men's and women's brains might account for the fact that far fewer women than men do well in science and engineering. Besides size, there do seem to be quite a few differences in the structure, chemistry, and function of male brains versus female brains, but no one has uncovered any evidence that these differences mean that women are intellectually inferior.

Some of those differences are there in your brain from birth, thanks to the sex hormones that bathe a fetus's brain in the womb. Researchers used to think that your sex affected only your hypothalamus, a small structure at the base of your brain that regulates hormone production, and basic mammalian behaviors such as mating, eating, and drinking. But today scientists know that your sex seems to affect your brain in many ways and therefore how you think, remember, and behave.

Information enters your brain in the form of sensory experiences—what you see, hear, taste, smell, and touch. Inside, brain cells called neurons form dendritic trees (multiarmed branching arrays of nerves), transmit this information to important parts of your brain, and receive signals from other neurons in return. Some parts of the brain are bigger, depending on whether you're a man or a woman—bigger in volume, that is, relative to the overall volume of your brain, which weighs about three pounds. For example, parts of the frontal cortex (which is involved in cognition), parts of the limbic cortex (which is involved in emotional responses), and the amygdala (a small almond-shaped structure that responds to emotion-triggering information), are all bigger in women. Some investigators even have found sex-based differences at the level of cells. In women, for example, neurons are particularly dense in parts of the temporal lobe cortex that process language and comprehension. Could this explain why women generally do better on language tests such as the verbal portion of the SATs or why they choose English as their college major more often than men do?

Other differences in brain structure between men and women also may explain why anxiety disorders are more common in girls than they are in boys; why men tend to find their way around by estimating distance, whereas women are more likely to use landmarks; and why men and women seem to learn and remember differently.

CHAPTER TWO

MEMORY: *SENIOR JEOPARDY!*, ANYONE?

If you've played the TV game show *Jeopardy!*, you've probably noticed that your response time has been slowing down over the years. That's because the brain changes with age. (That's why I believe that we should have a "senior" version of the game, where folks older than 50 get up to 10 minutes to answer each question, because they just can't seem to pull the answer—"I know it! Don't tell me . . ."—out of their memory banks.)

Forgetting things isn't all bad. (I'm not rationalizing just because I've always had a rotten memory. I've even forgotten names of people I'm interviewing—while I'm interviewing them!) In daily life, there are lots of things you don't need to remember—such as phone numbers you'll call only once or directions you'll use only one time. You wouldn't want to remember everything. That would be like saving all your junk mail.

Those dendritic trees make up different memory systems in your brain. When you experience something, the information enters your brain through different sensory pathways. The information fuses at some point into an integrated perception. When you remember something, your brain is stimulated to activate a certain pattern of neurons. Each time, that pattern is just a little bit different. Your memory doesn't work like a videotape or a DVD, where exactly the same experience is replicated each time you play it. Instead, your memory is constantly changing and evolving as you accumulate experiences and as your neurons grow and change.

So it's important to distinguish between memory *loss* (of which the major cause is Alzheimer's disease) and *change* in memory. Memory is very much a matter of perspective. Where you're standing, physically and emotionally, influences what you remember and how much. Consider how some memories are seared into your brain forever. For example, most people can tell you exactly where they were and what they were doing when they heard about the attacks on the Twin Towers and the Pentagon of September 11, 2001. In the same way, older people remember exactly what was happening around them when they heard that President Reagan had been shot, or President Kennedy had been assassinated. The emotions that accompanied those pieces of news are also why criminal investigators so often have to struggle with discrepancies among eyewitness accounts. Despite what you may have seen on *Law & Order* or any number of other television police procedurals, eyewitness accounts frequently vary dramatically—so much so that they are useless in convicting criminals. Eyewitness accounts may be inaccurate 40 to 60 percent of the time. Hard to base a life-and-death decision on percentages like that!

SENIOR MOMENTS

Why are all my friends and relatives who are over 40 convinced that they're losing their memory? It's actually quite rare to have anything seriously wrong with your memory. But memory does change. What happens as we age is that there is shrinkage inside the brain. The neurons themselves become smaller. What's affected is speed—how quickly the brain processes information. As the conduits—the neurons—grow smaller, the conductivity of neurons does slow down. Forgetting things as we grow older can be due to all kinds of reasons that have nothing to do with a serious illness such as Alzheimer's disease. Stress, for example, can be a culprit. So can illness, or remedies for serious illnesses. General anesthesia and chemotherapy can result in temporary memory loss, or "chemo brain." Heart and artery disease may be factors too if your brain is not getting enough circulation. But there's no single cause; memory changes are very individual and happen for different reasons in everyone.

Dr. Aaron P. Nelson, chief of neuropsychology at Brigham and Women's Hospital in Boston and assistant professor at Harvard Medical School, says that his patients bring him five complaints about their memories:

- Top of the list, what I suffer from: "name-nesia." You run into someone you know and you can't think of his or her name.
- The second most common: "room-nesia." You walk into a room and can't remember what you're supposed to be doing there.
- Third on the list is "episodic fleeting thought syndrome," or losing your train of thought. Suddenly you wonder: *What was I trying to say? Where was I?*
- Fourth is "parking-nesia," losing the car in the lot at the shopping mall.
- Fifth is "anavigationalism," a sudden spell, a moment when you miss the correct exit or the proper turn on the highway.

Dr. Nelson says he hears about these five types of memory lapses from as many as 90 percent of patients who seek him out. He emphasizes that none of these lapses means that anything significant is wrong with your memory. (I've added my own memory loss to the list: "key-nesia," where you can't remember where you left the house or car keys. I think it deserves its own category. I even have discovered a quick and easy cure for key-nesia. In your haste to drop the heavy groceries, you don't even look where you are dropping your car keys. Any place will do. So you never actually *see* where you left them. But if you watch where you put the keys, eye them carefully, you'll never lose them. What about that theory, Dr. Nelson?)

He says new technology can help people like me who have name-nesia. A handheld recording device is a "no-brainer," he says, to help you remember names. "You really shouldn't be relying on your brain for that kind of information. It's not what your brain was designed for. I'm totally reliant on my pocket PC. I consult it several times a day. For example, if I'm going to a social event, I list who'll be there and something about each person. It works! When I get to the party, that information is in the foreground."

THE GOOD AND BAD OF STRESS

Stress has a yin-and-yang effect on the brain. Too much stress can wreck your memory, says Nelson. "At very high and sustained levels, stress can cause actual physical and physiologic damage to structures in the brain that are important for memory. But at mild or moderate levels, stress actually can help memory function, believe it or not, because it kind of alerts you. It gets you to pay attention to something. It gets you to sort of label an event or an experience as important and worth remembering."

You should worry, however, if you can't remember information that is crucial to who you are, such as the name of your high school, "the name of a grandchild, some specific piece of information about where they worked for twenty years, that kind of a thing." Or if someone tells you something, and a moment later, you have absolutely no recollection of what's been said. That kind of memory lapse is more ominous and might spell real memory *loss*.

Nelson says that typically, people with more worrisome kinds of memory problems are *not* the ones who seek help. "It's usually some-one related to them—a spouse, a family member."

RX FOR YOUR MEMORY?

About 50 percent of how your brain ages is genetic. There's noth-ing you can do about it. But that leaves 50 percent over which you have some control. The higher your intelligence and the more in-tellectually inclined you are, the less likely you are to lose core memory. In 2001, an ongoing study of 678 aging nuns found that those who had written more complex sentences in their early twen-ties were less likely to have succumbed to senility. And in 2003, a study of London cabdrivers—people who have to find their way around for a living—discovered that their hippocampi were larger than normal.

If you want to stave off memory loss, you don't have to become a cabbie: Good nutrition and regular exercise are supposed to help, along with "use it or lose it"—the idea that you should stay mentally active. Some psychologists even give out "memory prescriptions," mental calisthenics or daily exercises—push-ups for the brain—or advise getting involved in the poker craze. (Seventy-one-year-old Doyle Brunson beat 666 opponents to win the 2004 Legends of Poker Tournament.) Others believe that mental activity should be-come a habit, something you cultivate so that learning new things permeates all your activities. Getting involved in things that engage you, going back to school, trying something new, taking up some-thing creative such as painting or jewelry design, playing games that involve planning and strategy, reading, joining debates and discus-sion groups—that's what your brain is for.

If you stay mentally active, you build up connectivity in the brain that can protect you against a siege of disease. Some nuns in the 2001 study had brain scans that revealed traces of Alzheimer's disease, but they showed no symptoms. Memory is about connec-tions, associating one stimulus with another. The denser the physical network, the "bushier" your dendritic trees, the more you can afford to lose before you show symptoms.

"What we've learned about memory echoes a lot of what we learned about cardiovascular health, in terms of brain health," says Nelson. "Everything that we know that's good for the heart is also good for the brain. So promoting optimal diet, exercise, managing stress—these things will all help ensure brain health as we age."

There's a lot of evidence, says Nelson, that ongoing intellectual activity may help stave off memory decay. "Whatever it is for you, something that engages your interests, something that engages your creativity, whether it's learning something new, getting involved in a project that puts you in contact with new learning, new people. These are all things that are going to promote that kind of lifelong intellectual engagement, which is important."

But what about cases like that of Dame Iris Murdoch? The famous British novelist's descent into Alzheimer's disease was the subject of the 2001 Oscar-winning film *Iris*. A prolific author of highly intellectual novels, Murdoch certainly had been mentally active. (As the movie illustrates, she exercised too: All her life, she loved to swim, even in old age.) Amazingly, a new study has found that daydreaming and woolgathering—something writers and teenagers are very prone to—may overwork and decay your hippocampus, the part of your brain most affected by Alzheimer's disease. So letting your brain idle—and not be focused on completing a task or reading or carrying on a conversation—could hasten the onset of Alzheimer's disease! On the plus side, the study may offer a way to diagnose Alzheimer's disease early, before symptoms become obvious.

Another possibility in Murdoch's case is that her mental acuity did give her greater resistance and she successfully held off her illness for years—until it finally overwhelmed her. Recent studies also have found that when people with a high degree of cognitive reserve are finally overcome by this disease, they fail very quickly. Nelson says there's been some interesting research out of Columbia University since the late 1980s on what's called "cognitive reserve." "Here, the notion is that being involved in lifelong learning and keeping intellectually engaged makes a difference. It certainly can push off the onset of Alzheimer's disease. We know that people with high levels of cognitive reserve can resist symptoms of Alzheimer's disease despite an equal or greater amount of brain pathology in the people who have the low levels of cognitive reserve. And this all has to do with both genetic and acquired intellectual interests."

As for nutrition, Nelson believes that there is good evidence of the importance of maintaining enough B vitamins in your diet. And if you can't get the vitamins from food, supplements are a good way to go. "One of the things we look for in patients presenting with memory complaints is what their nutritional status is, what they're eating, maybe what they're missing. It's important."

IS IT IN THE GENES?

I blame my name-nesia on my mother. She has had name recall problems for decades. But what about serious diseases, such as Alzheimer's disease? If any of your parents or grandparents are afflicted, does that mean you're going to automatically get it yourself? "Not at all," says Nelson. "The majority of Alzheimer's disease is sporadic. It doesn't have at least an obvious genetic connection, a familial connection. Certainly having a relative with Alzheimer's disease does increase your risk, but it's not anything like the type of genetic determination that we see with other illnesses. The genetics of Alzheimer's disease are extremely complex. We don't really have it worked out yet, but at this point, having a relative with Alzheimer's disease does not absolutely seal your destiny as having it yourself."

THE COLOR WHEEL

You may remember the color wheel from art class in grade school. You can find color wheels in art-supply stores, and craftspeople frequently rely on them when they work with colored paper, fabric, or beads. But you probably didn't know that the color wheel illustrates how the human eye and brain work. Early-twentieth-century German painter and color theorist Johannes Itten, a member of the famous Bauhaus School, invented the color wheel. Itten based his work on the research of Sir Isaac Newton and of physiologist Ewald Hering, who studied color blindness. Itten's color wheel is a visual tool made up of 12 colors. It starts with the 3 primary colors equidistant from one another: red, yellow, and blue. Mixing 2 primaries creates a secondary color, also equidistant from each other: purple, green, and orange. Mixing a primary color with a nearby secondary color creates a tertiary color. The 6 tertiaries are yellow-orange, red-orange, red-purple, blue-purple, blue-green, and yellow-green.

The color wheel helps demonstrate the phenomenon of "successive contrast," which happens when the brain creates an afterimage of a color when you look at it for an extended period of time. Green, for example, is the afterimage of red. (In Florida, you hear about the "flash of green" that supposedly you can see immediately after the sun sets in a blaze of red.) On Itten's color wheel, that afterimage is always the color's complement.

Try using the color wheel to come up with a split-complementary color scheme of your own choosing. Pick a key color. Then go directly across the color wheel to find its complement. Instead of the complement, use the two colors next to it. Itten recognized that the brain finds color equilibrium and recognizes harmonies. But of course, everyone sees color somewhat differently. Itten thought that his color wheel would help "liberate" painters and designers from "indecision and vacillating perception."

CHAPTER THREE

OLIVER SACKS: MUSIC MAKES THE MEMORY

Every disease is a musical problem.
—NOVALIS

You remember Oliver Sacks. Robin Williams portrayed him in the 1990 film *Awakenings*. He brings back to life a room full of hospitalized comatose patients suffering from Parkinson's disease by giving them a drug called L-dopa. Neurologist and famed author Oliver Sacks has spent a good deal of his life translating medical mysteries into common language, bringing science and medicine into the public arena in a way like no other storyteller. Through his musings and writings about his patients and his own life, Dr. Sacks has brought us greater understanding of neurological maladies, including autism, prodigies, vision loss, coma, and migraines and how they affect people in their daily lives.

As clinical professor of neurology at Albert Einstein College of Medicine and adjunct professor of psychiatry assigned to neurology at New York University School of Medicine, Sacks continues to discover and explore fascinating and mysterious workings of the brain. His latest project: music and memory, how music offers a unique way to look at the human brain, how the power of music transcends otherwise devastating memory loss to make life more bearable for people who are afflicted with memory loss.

First, some perspective. Sacks's interest in music and memory goes back decades, to 1966, to the patients portrayed in *Awakenings*. He discovered that before L-dopa was available as a treatment, his nurses reported that some of his patients would respond to music. "And then I rapidly saw this myself, that sometimes these people who could not take a step could dance. Some of these people who couldn't utter a syllable could sing. And that while they sang and while they danced, they were able to move freely. It was almost as if their Parkinsonism and their other neurological problems were bypassed. But the moment the music stopped, they stopped too."

Intrigued, Sacks called for a second opinion.

"I took the poet W. H. Auden along to the hospital to see this, and he quoted something of the German poet Novalis, who said, 'Every disease is a musical problem; every cure is a musical solution.' And certainly, one had the feeling that this was the case, although only for the few seconds or minutes that music lasted, with these patients.

"And then I was very struck. At some of the hospitals where I work, we have a lot of people with Alzheimer's disease and dementia." Sacks was struck by "how often how some of these people, although they'd lost the powers of thought and language, might recognize music, might sing along, might suddenly become lucid when they were exposed to music." It changed his views about the brain. "I think music perception, because it has such a widespread basis, is very robust, and even patients with widespread brain damage are usually able to respond to music."

A few years later, in the 1970s, Sacks saw a strange case of amnesia that made him think again about the power of music. "I saw a patient, one I call 'Jimmy,' who had lost so much memory he couldn't remember people or events, but he had a good memory for music."

Jimmy would be the first person Sacks met who could remember music quite well but only remember things he'd seen for just a few seconds. But Jimmy would not be the last case or the most extreme case like this. That would be Clive. Clive Waring.

Clive was an eminent musician and musicologist in England, an expert on Renaissance music, who in 1984 had devastating encephalitis due to herpes—shingles. This had the effect of destroying much of his memory; in particular, it destroyed what's sometimes called event memory, or episodic memory, so that he couldn't remember what had happened five seconds before.

The illness also deleted all the memories Clive had of the two or three decades before his encephalitis. At best, those days were but hazy apparitions. "He probably seemed to have no reliable memories since childhood. He didn't recognize the room in which he was. And so it came as a great surprise, given this utter devastation, when it turned out that his memory for music and his ability to sing, to play the organ or piano, to conduct an orchestra or choir, was brilliant and quite unimpaired."

Even with devastating brain damage, Clive could sit down at the piano and play a whole piece of music from beginning to end. He could stand before musicians and conduct an entire orchestra "and beautifully, turning to all the different sections of the orchestra and, obviously, knowing the piece in great detail. When he conducts an orchestra or a choir, when he sings, he seems absolutely normal. You wouldn't think that anything was the matter. But five or ten seconds after he's finished, he'll have no recollection of it. And sometimes if you ask him, 'Do you know such-and-such a piece?' or 'Do you know this particular Bach prelude and fugue?' he'll say no. But if you start him on one note, then he's got it. So he's not entirely conscious of what he has."

Clive will also forget faces. A few seconds after he meets you, he will not remember having met you.

"You might not realize this at first from the charming way in which he greets people. He improvises his greetings, but he doesn't genuinely recognize anybody except his wife. The two of them were married a little bit before his encephalitis, and they're still very much in love.

Clive does not recognize his wife, though, when she walks by. But he does remember her "sound."

"He recognizes her footsteps. He recognizes her voice. He recognizes her gestures, her approach. And above all he recognizes her kisses. So he recognizes sort of physical and auditory contact with his wife."

Sacks says Clive has retained other small vestiges of normality. He is able to dress himself "very nattily and to talk fluently and to walk around" and perform all sorts of various skills. But it's Clive's preserved musical powers that very much interest Sacks.

"People sometimes use a different term and call this procedural memory. But whether this rather narrow-sounding term can be expanded to include all of his musical powers, I don't know."

Sacks says he is interested in other memory-musical phenomena such as musical hallucinations, "where sometimes relatively unmusical people will hear something in great detail, a detail they never knew that they knew. Typically in musical hallucinations, the music hallucinated is usually music from early life, sometimes very, very early life. I begin to get the impression that musical memory can be extremely exact and faithful in pitch, in time, and all the subtleties in almost everybody."

Sacks says that while most of Clive's musical skills were learned before his brain was injured—"it's more a preservation of skills and knowledge, musical knowledge, than any enlargement of them"—he thinks that Clive and some other people with amnesia can continue to learn.

"Some years ago, in my *Anthropologist on Mars* book, I'd described an amnesic young man musically inclined called Greg and how on one occasion I took him along to the Grateful Dead. He'd been a Deadhead in the 1960s. He sang along with the earlier songs. He was very puzzled by the later songs. He said it's like the music of the future, which in a way it was for him. Later he had no recollection of having been to Madison Square Garden, but he did have some recollection of the new music. And so I think that new music can be learned by the amnesic."

Sacks has known about Clive since the mid-1980s. Most of the research about Clive's unique musical abilities involve merely observing him as he goes about his life. But Sacks believes just watching him and taking notes is not scientific enough.

"He also needs other sorts of studies like brain imagery to show exactly what is going on in his brain when he perceives music, when he recollects music, when he plays music. The techniques like this weren't available twenty years ago."

Sacks uses techniques common to studying brains today—brain scanners—which can observe the parts of the brain as they become activated by music. He's found that music is widely infused throughout the brain.

"So many parts of the brain light up when music is being played or imagined or hallucinated or recollected. It's not only in many areas of the cortex on both sides." He noticed that the music energizes the cerebellum and the basal ganglia—near the brain stem—which, he says, may be involved in the sense of rhythm and timing, "which, for example, is knocked out in Parkinson's disease, which is why music is so important for people like that."

Whether there's any sort of final music area, the way there is a language area, people are not certain. But there may be one in one of the temporal lobes.

As a clinician, Sacks continues to study the surprising effects that music has on the brain. He has met people in whom seizures have been elicited by music.

"I saw a lady who was originally found unconscious with a bitten tongue by the radio. The last thing she could remember was that there had been Neapolitan songs on the radio. No attention was paid to this but then when this occurred again and again, it became clear that Neapolitan songs—and no other music—were a specific trigger for her, and again this must tell one something about the brain.

"I did recently write about a patient, a woman, who had become aphasic—who had lost language—but had not lost her musical abilities and how sometimes she could get the words of a song, not only the music of a song but also the words of a song. This at least pleased her very much because it showed that the language was there somewhere, whether or not it could be disembedded from the song."

MUSIC THERAPY

If music has the power to evoke such hidden talents in brain-damaged people, perhaps it has healing powers too. Many physicians, says Sacks, use music as therapy.

"In 1973, when a documentary of *Awakenings* was made, when the film director came to our hospital, he said, 'Can I meet the music therapist? She seems to be the most important person around here.' At one of the hospitals where I work, we have an institute for music and neurological function. And I think music therapists are very important, both in practice and as researchers and opening doors, and I have great respect for them."

One region ripe for research, says Sacks, is the brain's frontal lobes, the creative, inquisitive part of the brain. Music may offer a window inside. We are only just beginning to study creativity and what goes on in the brain, says Sacks. "Certainly the improvising musician is the nicest, and, in a way, simplest and most accessible example of creativity at work, and I agree it would be lovely to do such studies.

"So I think of music as an unexpected way of getting into the mind and brain, in all sorts of ways, in all sorts of directions. And it's relatively new. If you look at a neurology or neuroscience book of twenty, twenty-five years ago, you don't find any reference to music."

Yet music, says Sacks, is "as rich as language. I think we are a musical species, no less than a language species," but the musical side of our species has been ignored, he says, "or treated anecdotally" because we don't think it's important to survival of the species.

"Certainly, Steven Pinker [cognitive psychologist at Harvard] has referred to music as auditory cheesecake—or just something which happens. And I suspect, as some others do, that the origins of music and speech go together, and the two of them coevolved. There's certainly no culture in which music isn't important, whether as a means of expression or communication or ritual or enjoyment."

CHAPTER FOUR

PATHWAYS OF ADDICTION

He raised his eyes languidly from the old black-letter volume which he had opened. "It is cocaine," he said, "a seven-per-cent solution. Would you care to try it?"

—SHERLOCK HOLMES,

The Sign of the Four

If you look back in history, whenever we didn't understand a disease or an illness, our response to people who got sick was always the same: We blamed the person with the illness. There was a time when people who had cancer were stigmatized. No one would dare say the c-word out loud. The same was true, and still is, to a certain extent, for mental illness, such as depression.

But as we become more enlightened about disease and disease processes, we have come to understand that cancer or mental illness is not a moral failure or a character weakness but a disease. The same can be said about our attitudes toward addiction. Only now are we beginning to understand that addiction is a disease—that there are real biophysical changes that occur in the brain that can make some people vulnerable to becoming addicted and make it nearly impossible for them to quit; time after time, they relapse.

"There is no question that addiction is a disease," says Dr. Nora Volkow, director of the National Institute on Drug Abuse, part of the National Institutes of Health. "Research has shown that it affects the brain in very specific ways that can help us understand why, through this damage, the person loses control of his or her action, as it relates to taking drugs. Despite this, it is still not accepted as a disease by most people."

Even physicians, at times, have trouble. "Unfortunately, yes, indeed, even physicians do not recognize that sometimes it's a disease." One of the problems is understanding when a habit becomes an addiction, when it crosses the line from just being an annoying trait to a life-threatening one. "We have a number of things that we look for, in terms of when we begin to call something an addiction," says Dr. Shelly Greenfield, associate professor of psychiatry at Harvard Medical School and associate clinical director of the McLean Hospital Alcohol and Drug Abuse Treatment Program, in Boston. "We're really looking for adverse consequences in the person's life, such as a person spending more time using the substance, or using it in greater quantities than they want to, giving up activities because they're using the substance or recovering from it. They continue to use something in spite of the fact that they know it jeopardizes their physical or their mental health."

SPOTTING THE SIGNS

But how easy is it for physicians, nurses, and other health care providers to spot signs of substance abuse and addiction? Greenfield says physicians and nurses are often not trained well enough to see evidence that might be right in front of them.

"In fact drug and alcohol addiction are among the largest, most prevalent problems that we have; they present a major public health problem in addition to an individual and family problem. And at almost every health care setting, whether it's a physical health care setting or mental health care setting, patients present themselves for other types of disorders and they also have a co-occurring substance use disorder." This presents an opportunity for physicians and nurses to actually engage and screen patients. Are they using alcohol or drugs in a harmful way? Have they already become addicted or dependent on these substances? "A doctor can actually catch a patient, where something is moving forward into becoming a major abuse or dependence problem, and can help by educating a patient and referring them to treatment, to intervening earlier on in the course of the disease process. Of course, in every area of medicine, what we try to do is intervene as early as possible. But often doctors and nurses haven't had adequate clinical training to screen and diagnose and to understand what the treatment processes are."

In many cases, says Volkow, doctors in emergency departments are afraid to ask patients if they are addicted to drugs. "Why don't they ask the question? Because drug addicts are stigmatized and doctors feel uncomfortable asking the patients do they drink, do they take cocaine. And they don't even know how to ask that question sometimes—and certainly they don't recognize it." And even if they do know to ask the right questions, they may not know how to follow them up. "Unfortunately, there is no parity for the treatment of drug addiction. So as a result of that, many medical insurances will not cover the cost of doing an evaluation for drug addiction and proper referral. A patient that is addicted when they have a job or they are referred by their physician is very different from the situation of a person that is homeless, that doesn't have a job, that doesn't have a family, that ends up in the emergency room, and you are actually hand-tied in terms of what you can offer."

And many times, people with addiction problems may never show up in a physician's office or emergency department. Though they appear in countless television and film plots, the people who continually show up in clinics for treatment represent just the tip of the iceberg, says Greenfield. "In fact, the vast majority of folks who have substance abuse disorder problems, many are working every day. They are taking care of various kinds of responsibilities out in the world, and they are actually having problems with addiction. Sometimes those things manifest themselves at home rather than at work. Sometimes they may not show up in an emergency room, but they may show up, maybe, in a mental health care clinic or a physical health care clinic.

"So this is not just in the emergency room with someone who's coming in acutely, multiple times, but it's much more generalizable to all sorts of health care settings where we, as medical professionals, can do a much, much, much better job at diagnosing early and referring for treatment many, many individuals who could actually benefit from all the available treatments that do exist and are actually effective."

"THIS IS YOUR BRAIN ON DRUGS . . ."

Remember that commercial? Really an anticommercial, about the dangers of drugs? The frying pan, the burned egg? It sent a clear message that addiction is not just a bad habit but is also a complex chemical interaction in your brain. And one of the breakthroughs in the addiction field was the realization, says Dr. Rob Malenka, professor in psychiatry and behavioral sciences at Stanford University, that whether it's nicotine or alcohol or cocaine or heroin, they all work on the brain's reward circuitry.

"Through evolution, the brain has evolved to tell us what feels really good, what is rewarding, what is important for our survival. And we now know that all these different drugs of abuse act on this specific circuit in these specific brain areas." One of the key chemical messengers in the brain circuits is a substance called dopamine. "It's a substance we term a neurotransmitter, and it turns out that all these different addictive substances increase the actions or the release of dopamine." And while the effects and actions of dopamine are just now being understood, the release of dopamine in certain brain structures tells the person that this substance is reinforcing or rewarding.

"And then, for certain genetically vulnerable individuals, there are long-lasting changes in these circuits that lead the person to believe that the pursuit of this substance is the most important thing in their life." These long-lasting changes occur in the connections between nerve cells, called synapses. "So the communication between individual nerve cells that are part of this circuit starts to change. There are molecular changes in these cells that are part of these circuits. We're beginning to learn a reasonable amount about what are these changes in specific connections between nerve cells. And that's the first step towards trying to understand how to reverse those changes."

Because reversal is the key to drug addiction. As Mark Twain once said about his own addiction, "I don't have much trouble giving up smoking. I've done it a hundred times." In many cases it's possible to stop the addiction, give up the cigarettes or cocaine. But what happens is that people go back to smoking or snorting. They can't stay away. Once those chemical and neurological changes take place in the brain, reversing them is not very easy. The addiction has rewired the brain and, very importantly, brought into that rewiring the part of the brain that encodes memories, so that a relapse may occur without the person even being exposed to the addictive substance but simply to the *memory* of that exposure.

"It's extraordinarily important," says Volkow, "in the terms of why it's so difficult to treat addiction and why people, despite the fact that they face catastrophic consequences—not negative, catastrophic—and they don't want to take the drug anymore, they relapse. It's almost like a reflex." Volkow is very clear and determined on this important point. She wants to make sure you understand just how difficult it is for someone to *not* relapse when exposed to that memory. It's almost like uncontrolled salivating when you think about a great dessert.

"Inside your brain, there is a release of dopamine when the person that's addicted sees stimuli associated with the drug that activates the motivational circuit almost in a reflexlike way. And that drives him or her to do that behavior. And that's evidently one of the mechanisms why relapse occurs and it's so difficult to 'kick the habit.'"

So finding a way to erase that emotional circuitry is one of the great challenges. "Indeed, that's one of the strategies that we're now trying to encourage investigators to look at: the development of medications that can either erase those memories associated with the drug or, alternatively—very important—can create stronger memories that can overcome those learned responses. So that your behavior is not driven by what we call conditioning, but by these new learned experiences."

There is ongoing research to erase those neurological pathways, but so far only in lab animals. "But there are some real interesting positive results that suggest that this strategy may, in fact, prove beneficial in helping people through the therapeutic process."

But what about other addictions that do not start out with well-known street drugs such as nicotine or alcohol but instead with addictive behavior about activities such as gambling, eating, or playing video games. Is the brain laying down the same kind of new pathways? Malenka says the general consensus among scientists is that "yes, a lot of these other kinds of compulsive, especially rewarding behaviors, or reinforcing behaviors like gambling, like overeating, like perhaps even video game playing, certainly affect these so-called reinforcement reward circuits, they do affect the release of this chemical messenger dopamine. Work done by Volkow and others has shown that it's not only the release of dopamine but also how *much* is released and how *fast* it's released that is important.

"And it turns out, the highly addicting substances, like cocaine, can really cause a much more rapid, stronger increase in this chemical messenger than, for instance, what I do all the time, which is eat doughnuts, or eat a quart of Häagen-Dazs ice cream, which is highly rewarding for me. But I can kick the habit when I choose to."

WATCHING THE CRAVING IN THE BRAIN

Volkow agrees. She says it's a "very interesting question that has started to intrigue many of the scientists; certainly, it has intrigued me for many years." Using brain imaging, Volkow and her colleagues at the Brookhaven National Laboratory in Upton, New York, have watched the brain in action as it reacts to different stimuli, "specifically in pathological eating in obesity, versus those that we see in addiction." And what she sees are both similarities and differences. The similarities: there is a marked disruption of the functioning of the dopamine system, which is directly affected by drugs, "but it's also the one that motivates our behavior vis-à-vis natural activities like eating or doing social interactions or engaging in procreation—sexual behaviors."

The dopamine system becomes dysfunctional, she says, in both addiction and overeating. The dopamine is not released in as great a quantity as it was before, so that it does not produce that terrific sense of well-being—the high—as it used to. "And it is believed that one of the reasons why there is a motivation to either continue taking the drug or to compulsively eat is that it's a mechanism to compensate for this deficit." In other words, you eat more or take more drugs to stimulate more dopamine release. It takes more to achieve the "high."

On the other hand, what's uniquely different in this case—eating versus drugs—in pathologically obese people "is that their brain is particularly sensitive to the pleasurable aspects associated with food. Evidently, that is the reason why they are favoring these particular stimuli—in this case, food—over other ones. So this is why some people become addicted specifically to a certain substance and others may become addicted to behavior, because each one of our brains responds with a different sensitivity to the rewarding effects of the stimuli."

THIS IS YOUR BRAIN UNDER STRESS . . .

It's well known that stress can make people relapse: *I need that drink, just one bite of cheesecake.* What link is there to addiction? "It turns out that the circuits in the brain that respond to stress," says Malenka, "that release certain hormones in response to stress, are heavily interconnected with the exact circuits we've been talking about, the so-called reward circuits, the circuits that use dopamine. And work from many labs has shown that in humans as well as in animal models of addiction, stress is a very important factor in causing the continued use of a substance, as well as leading to relapse." It appears that "the brain's response to stress is actually pretty similar, under certain cases, to the brain's response to certain drugs of abuse."

For example, in a classic set of experiments it's been shown that "if you train an animal to self-administer a drug—this happens in human beings too—and then you take the drug away for many weeks or months, an acute stressful event can have that person or animal start using the drug again. We believe that's because that stress is causing perhaps the same release of dopamine" that substance abuse causes.

Volkow points out that there is an overlap in the circuits and brain areas that are affected by drugs and stress. And what ties those areas together? What commonality do they have? It all boils down to dopamine. "Dopamine is there not only to signal pleasure but actually to signal saliency, and as you recognize, of course, pleasurable events are very salient." We need to learn from them. "But a stress is also very salient, because if we do not learn from it, when we are exposed to it again we may not avoid it.

"So anything that has importance, in terms of the survival of the species, that connotes a need to learn an experience, so you can change your behavior accordingly, will involve dopamine. When you are exposed to a stress, you are going to be releasing dopamine. And that's going to drive a similar circuitry to that that we see in drug addiction."

In individuals who are addicted, the increase in dopamine by itself is a conditioned response. "It's a learned memory response that's associated with the drug. And this is probably one of the reasons why, when a person that's going through recovery is stressed, and then they relapse, it's in a way similar to the way they relapse when they get exposed to a stimulant that, in the past, they had associated with the use of drugs."

STIGMA: ROADBLOCK TO RESEARCH

Is there any way, medically, to reverse the addictive wiring in the brain, perhaps through new medications? Volkow says it's a "very challenging question indeed." The federal government is trying to encourage researchers using laboratory animals "to do exactly that. Can you strengthen certain pathways that have been damaged by the chronic use of drugs? There are some very interesting compounds that, for example, are targeting the disruptions that exist in the memory circuit. We're also looking to strengthening the ability of your brain, through cognitive operations, to regulate your emotions and your desires. And this is, of course, a pathway that is badly eroded by the usage of drugs. Those compounds are currently being investigated in animal studies, some with actually fascinating positive results. But it's very difficult to get these compounds into the clinic."

Why can't we move toward testing these drugs in humans? First, says Volkow, because the process is very, very expensive, and as is the norm in developing new drugs, that burden falls on the pharmaceutical industry, "and here in the pharmaceutical industry, it's not one of their primary interests." It's not just the expense that turns them away. Drug companies routinely spend hundreds of millions of dollars developing a new drug. In this case, Volkow believes the reason is quite different. Drug companies fear "that drug addiction is stigmatized." And drug companies don't want to be associated with unpopular illnesses. "That certainly doesn't help the translation of potentially promising medications into the practice. We cannot get them into the clinics. We cannot get that translation from the animal experiments into the humans as fast as we could, because of restriction and budgets. The government has to carry the costs that are associated with these medication developments."

What may change this attitude is the profit motive. Drug companies understand profits. Drug addicts may not constitute a large enough market to be profitable, especially if many of them are not covered by health insurance. But what is slowly dawning on researchers and the drug industry—with the prodding of Volkow and researchers like her—is the potential for much larger profits from a much larger pool of customers. "You don't necessarily need to address it as a medication for cocaine addiction but rather a medication for addiction in general. And then you can start to recognize that, indeed, the market could be very large because some of these processes also underline some of the behavior of compulsive disorders, such as pathological gambling or compulsive eating. These medications could also be beneficial in addressing some of the disruption that we see in these individuals." Imagine a drug that stops you from overeating. Or treats gambling. Just think of the numbers of potential customers. Tens of millions. Hundreds of millions. Billions, worldwide. Even on a smaller scale, says Volkow, "the ability to develop a medication to treat methamphetamine addiction would be quite extraordinary."

TREATING ADDICTION LIKE OTHER CHRONIC DISEASES

Until these new brain drugs are developed, therapists will treat patients the best way they can with the drugs currently available. Dr. Greenfield stresses that understanding the neurological changes in the brain can help explain to patients and their physicians why getting well and into recovery can be so difficult. "On the other hand, the other part that's so important for people to understand is, like other types of medical disorders, addiction is a treatable disorder and people do actually get better. But like other medical disorders, like diabetes, hypertension, heart disease, people generally don't get better with a single treatment. They often require several different treatments, sometimes different types of treatments over time, to slowly but surely regain their health—their physical health and their mental health.

Greenfield believes that people have an outdated view of addiction and its treatments. "This stigmatization has really held back the general public's understanding of all the gains that have been made over the last 20 years: People actually get better; people do much better. It's a classic 'bench to bedside' story: basic biology and treatment-related research demonstrating over and over again that many, many patients can be helped. We are able to treat a vast majority of folks, and in many instances, have people who are in recovery for many years, sometimes with relapses that occur periodically. Hopefully, if they remain connected to treatment and to a treatment community, they can recognize a relapse quickly and associate themselves again with treatment that's been helpful and shorten the actual duration of their actual relapse or lapse.

"And that's what we aim for: to keep people as healthy as we can over time, and if they do slip—a relapse—to shorten the duration of that and to return them as quickly as possible to their best-functioning selves."

The treatments may not be solely through drugs, says Volkow. "There's actually behavioral cognitive group therapies that have been unequivocally shown that they work. They are effective in the treatment of drug addiction, and yet they are not necessarily all the time accepted as such." What's not accepted or understood by the general public is that addiction should be looked at and treated like other chronic diseases.

"One of the reasons has been that people expect the person that goes to treatment to be miraculously cured after going through a rehabilitation program. So they go through the rehabilitation program, they stop taking the drugs. Six months later something happens in their lives and they relapse, and that is then used as an argument. 'You see? Treatment does not work!' Of course, we would never use that argument for someone being treated with antihypertensive medication for high blood pressure. The moment he or she stops taking the medicine, blood pressure goes up; you'll never use the argument 'You see, that does not work.'

"This relates to the whole stigmatization aspect of it, very differently from the way that we treat other diseases, in terms of what we expect of the treatment of drug addiction. We expect a cure. Yet we know that it's a chronic disease. So we are treating. Very rarely do we cure right now."

Greenfield cannot stress that point enough. "If you follow along on that model and continue to think about it—for someone who's being treated for diabetes or heart disease or hypertension, what you hope you would do is you would help them in the acute first illness that they had, and then you enter them into treatment. And you want them to be monitored over the long term to keep them healthy and hopefully to prevent any more acute episodes.

"This is very similar to substance abuse treatment. People come in with an acute problem, you treat them, and then what you'd really like to do is enter them into care where they're followed to keep them as healthy as possible. If people are connected to care, if there is a stressful life event or something else that comes up, they're already hooked into care and they can usually continue how they're doing or circumvent any exacerbation in their condition, hopefully avoiding, if possible, another major episode.

"This is truly a model of keeping people well over time, one that it's really important we begin to implement for people with substance use disorders. We have many models in chronic disease care that work. We know that this is a kind of model that will work for substance abuse as well, over the long term."

LOOKING AHEAD

Researchers are always looking for new questions to answer. What would we like to know about the brain that we don't know now, and how could we get to know it? Where should the limited dollars for research be invested?

"I would put more money into really defining both the molecular changes in the brain that happen in response to substances of abuse or during addiction, but really beef up the imaging approaches so we can really understand, because the devil is always in the details," says Dr. Malenka. "Which specific brain areas, which specific connections, which circuits in the brain are really being modified in a semipermanent way during addiction?"

Malenka also thinks we should invest in trying to identify both genetically and through other approaches those people who're going to be most vulnerable to developing addiction, especially kids. "I believe people really start developing their problems mostly during adolescence, late adolescence to early adulthood, and I think it would be wonderful if we could identify those specific individuals that are going to be particularly susceptible and vulnerable to addictions."

As for Volkow? "I'd like to be able, with the money, to create the knowledge that would allow us to have devoutness for substance abuse—the knowledge that would drive and motivate and intensify the pharmaceutical industry to be able to fund medications into clinical practice. And finally, to use that knowledge to create more targeted prevention, such that less people get exposed to drugs."

CHAPTER FIVE

SLEEP AND LEARNING: CAFFEINE IN YOUR BEER

*Laugh and the world laughs with you, snore and you sleep
alone.*

— ANTHONY BURGESS

Birds do it. Bees don't. We do it more than elephants but less than
bats or opossums. Of course, I'm talking about sleep—something you
can't do without, yet few of us, it appears, get enough of it.

Writers, poets, and scientists have pondered this mysterious state
of consciousness for centuries, asking questions such as: Why do we
even need to sleep? Why would evolution favor sleep? Think about
it. When you're asleep you're in your most vulnerable position. You're
unable to defend yourself from an enemy. So what's the payoff here?
Are there crucial biological functions that can take place only when
we sleep? And if so, what happens when you don't get enough sleep?
Does memory suffer? Are there deep psychological things that are
going on when we sleep that we don't even know about?

Yet millions of people are not getting enough sleep. Whether it's because of snoring, restless leg syndrome, or sleepwalking with a pint of ice cream, a staggering number of people don't sleep well at night.

SO WHY DO WE NEED TO SLEEP?

Sleep is much more important than serving as a simple mind refresher and "cobweb" clearer. Researchers are now able to peer into the brain and watch what happens when we sleep. They are finding that sleep is quite necessary if we are to learn, remember, solidify, and improve new skills. And a good night's sleep may also be necessary for repairing an injured brain.

First, to debunk a myth: when you go to sleep, while you may be unconscious lying there in bed, your brain is far from being "asleep."

"There are probably many, many things that are occurring during sleep that are beneficial, things that have been assigned by evolution to that part of the night," says Dr. Robert Stickgold, associate professor of psychiatry at Harvard Medical School.

"My own personal favorite, and the one that I think has the best explanation for why it would have evolved originally, is that while we're asleep the brain is going through our memory stores and trying to see what's worth keeping, what's worth throwing out, and how things should be strengthened and put together."

That doesn't mean that our bodies are not using sleep to replenish and refresh us. The bottom line on sleep, says Dr. Carlos Schenck, senior staff physician at the Minnesota Regional Sleep Disorders Center in Minneapolis, is how well you feel in the morning.

"If someone says, 'I don't think I'm getting enough sleep,' and yet they wake up refreshed and they function, they may be a short sleeper and may not need nearly as much sleep as the average person. So you always have to focus on the functional consequence."

Judging how much sleep you need is very difficult. But research now shows that if you can function well during the day, and people tell you that you're looking well and not drowsy, then you're probably getting enough sleep.

But "if you need a double grande to get that morning going, then you're probably not getting enough sleep," says Stickgold, referring, of course to the caffeine jolt of coffee.

"And there's another ominous event that's taking place now," warns Schenck. "There are new beers on the market that are loaded with caffeine, and so people will drink a beer in the evening and the caffeine makes them feel more alert, and yet it will greatly prolong their onset for sleep, and that could have a devastating effect on their sleep–wakefulness rhythm. You'll be hearing more and more about these types of beers loaded with caffeine. I think it's really going to have a very devastating effect."

CEMENTING NEW SKILLS

But getting back to that memory consolidation idea, where it appears that our minds are using sleep to sort through what we learned during the day, Stickgold and other scientists have conducted tests that show just how crucial a good night's sleep is to learning new skills.

Scientists noticed that when they trained study subjects on a new task and then tested them on the task a few hours later, they didn't show any improvement at all. What they did find is that the improvement would show up later—the next day!

They discovered that people really do need to sleep on it to cement their new learning experience.

"Even if we trained them right before they went to bed, the next day they show improvement," says Stickgold. "And it's pretty consistent from person to person. But it varies with how much sleep they got. And the thing we saw that I scare my college students with is that the study subjects who had less than six hours of sleep showed no improvement at all the next day."

How many college kids get six hours' sleep?

"Although when you talk to those students who cram for exams, they'll tell you, 'I did fine,' the next day. But two days later, it's all gone. And that might be a sleep connection. We don't know that yet. But for this particular test, the more sleep you had in excess of six hours, the better you performed the next day."

And what's going on in your brain that you need this sleep to cement that practice? Stickgold took his students into his sleep lab to find out.

"We got very surprising results."

What he found is that when you go through a night of sleep you go through a 90-minute sleep cycle. You go down into deep sleep and then come out of it into REM sleep, where you do most of your intense dreaming. This cycle gets repeated all night.

"And what we discovered is that to show improvement on the next day, you need good slow-wave—or deep—sleep in the first couple of hours of the night. And you need REM sleep, but only after six hours. And that's probably why those students who got less than six hours didn't show improvement.

"So you need two things to happen at night. You need to have something happen while you're in deep sleep, earlier in the night. There's three or four hours that have to pass. I don't know what's happening. I describe it as 'the dough is rising.' And then at the end of the night, you need REM sleep, where presumably something else happens. And that's what finally cements it all together."

"A night of sleep," says Dr. Matthew Walker of Harvard Medical School, "reorganizes the representation of a memory within the human brain, making memory more efficient." In a test, Walker taught 12 healthy college-aged study subjects a finger-tapping test similar to piano lessons. The study subjects were retested either after 12 hours containing sleep or 12 hours of awake time. During the retesting, the brains of the study subjects were scanned by fMRI and their brain activity was watched. And the results were clear.

"After sleep, you improve your performance by about twenty percent to thirty percent," says Walker. "Without sleep, there is no improvement. So practice with a good night's sleep makes perfect."

And the idea that you need a good night's sleep to nail a new skill is quite evident when you talk to people who are practicing new skills, such as athletes and dancers.

"A gymnast tries and tries a certain move, and in total frustration after an hour just gets off the beam and says, 'I'm going home.' They come back the next day and they get on the beam, and they've got it. And it's sort of magic, or they say, 'Well, I must have been tired.' But if you talk to them—if you asked them, 'Well, if you practiced in the morning, would you come back in the afternoon?' They'd say, 'No, I'd wait until the next day.' So for those sorts of skills, whether it's a pianist stuck on a passage who finally has to put it away, or a gymnast, that, I think, is very likely to involve this kind of sleep stage requirement."

Stickgold says that it doesn't matter at what time during the day you do your training, as long as you get six-plus hours of sleep at night. Which explains why, as we get older, it gets harder to learn new skills—older people get less sleep. Maybe nature knows this. Maybe when we are young, our bodies know we need more sleep to learn new skills, so we sleep longer to consolidate them.

"Everybody knows that when you drop below six hours of sleep, you're not running on all cylinders. And it's just trying to fight a strange sort of macho culture thing of sleep deprivation being cool. And it's a bad scene throughout the country, the amount of sleep deprivation. There are estimates that there might be more traffic fatalities from fatigue than even from alcohol."

There is a study showing that immediately on waking up, people's reaction times and reasoning times are equivalent to those when they're legally drunk. You don't want to jump out of bed and run an errand in the car. But if you're a medical resident at a hospital on call for 24 or more hours, you have no choice.

"I think we want to say, 'Physician, heal thyself.' I think it's a big problem. People seem to know it and seem not to want to deal with it. There was a wonderful study with students, where they would wake them up in the middle of the night—math majors—and ask them to do math problems. And they would write out the solutions to these math problems and go back to bed. And when they checked them afterwards, they discovered that for the first line or two, they'd be doing fine, and then it would just be garbage. I think there's a real issue there for how much they're learning. But even more important, when you wake someone up from deep sleep and you look at their EEG—their EEG says their brain is still asleep—and I think some of these residents are prescribing while they're still asleep."

These new findings about sleep and consolidation of skills may explain why teenagers and babies sleep so much. Dr. Walker says that infants' brains are continually learning new motor skills, which may demand a great deal of sleep to consolidate. As for teens, they may be playing sports, learning to play an instrument, or learning to drive a car—all practiced skills that need sleep to be cemented into the brain's wiring.

Injured brains may benefit from sleep too. "A good night of sleep may be able to help reestablish connections in the brain of stroke victims," says Walker. "Since the brain undergoes these plastic changes, patients can take on new tasks and learn new ways of doing things they did prior to their stroke. Sleep may incrementally assist their recovery."

Walker and his colleagues plan on taking this next step: testing this theory with stroke victims to see if their condition improves with sleep.

So to learn a new skill, to perfect that new Mozart concerto, nothing may help more than a good night's sleep.

PART II

—

COSMOLOGY

CHAPTER SIX

WHERE THE VERY BIG MEETS THE VERY LITTLE

There is a theory which states that if ever anybody discovers exactly what the Universe is for and why it is here, it will instantly disappear and be replaced by something even more bizarre and inexplicable. There is another theory which states that this has already happened.

— *The Hitchhiker's Guide to the Galaxy*

Isn't the universe supposed to be getting easier to understand? I mean, aren't all the smart people who study quantum physics, relativity, string theory, extra dimensions, and water on Mars and bring back samples from comets encased in tennis racket–like foam collectors supposed to be making the world look simpler? If so, then why does it appear that the more we know, the more we don't know? The more we learn, the more there is to learn? Still?

I mean here it is the twenty-first century. A hundred years since Einstein was in his heyday, almost four decades since humans set foot on the moon, 50 years since *Sputnik*, not to mention the *Voyager* missions to the planets and telescopes, such as the Hubble Space Telescope, that can peer 13 billion years back into time. It is the age of leptons, baryons, muons, neutrinos, and antimatter. Yet we still

51

have no idea what 96 percent of the universe is made of. (And oh, by the way, we didn't know that we didn't know what most of the universe was made of until very recently.) And we don't know why there shouldn't be *more* of the stuff we don't know about, either.

It's very humbling. But at the same time, very exciting. "It's the golden age of physics. I wouldn't want to live in any other time." You hear that a lot from astronomers, theoretical physicists, and mathematicians whose numbers cover blackboards across the world. There is no better time to be present at the future than in the wild and woolly world of physics.

On the other hand, the more physicists discover about the universe, the stranger it appears. "I don't see much indication of it getting terribly simple just yet," says Dr. Roger Penrose, professor of mathematics at Oxford. "I think one likes to think that eventually there will be some simple principle that governs everything, but we're certainly a long way from that now."

THE THEORY OF EVERYTHING

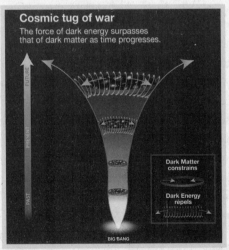

The timeline shows how dark matter was the prevalent force in early universe, with dark energy overtaking it. NASA, ESA, and A. Feild (STScI).

To understand the excitement, we've got to go back a hundred years or so. Back at the turn of the century—that one before this one—a lowly patent office clerk, Albert Einstein, turned the world of science on its head, in an era that became known as the age of relativity (1905–1915). In 1905, he published his special theory of relativity, which equated mass with energy: $e = mc^2$. Ten years later, he would publish his general theory of relativity, which described gravity as nothing more than a curvature of space, caused by the presence of massive bodies such as stars and planets. Einstein was so far ahead of his time that some people considered his 1905 theory to be quite nutty. The 26-year-old physicist couldn't even secure a teaching job.

Nevertheless, his seemingly oddball ideas about the universe—such as the path of light being bent by gravity—were verified in 1919 by observations made during a solar eclipse. Thus, the predictions Einstein had made in advance about his theories were borne out by experiments. (That's a very important concept for a theory to gain acceptance: the ability to conduct experiments to test predictions made by the theory.)

What followed was another sharp turn in science: the age of quantum mechanics. Instead of looking out into the vast cosmos where galaxies and stars warp space, quantum mechanics looked in the opposite direction at the ultrasmall world of the atom. It found that movements of subatomic particles, such as electrons around the nucleus, could not be precisely described the way planets orbiting the sun can be. Unlike the well-defined, smooth orbits and flowing curves described by Einstein's theory of relativity in the wide open spaces of the universe, the paths of the subatomic particles followed a whole other set of rules, in which you could not really ever be sure where they were. You could say they were probably here or probably there—and give the odds of finding them in any of those places—but you couldn't know for sure. And yet in this tiny subatomic world, the probabilistic math of quantum theory proved to be extremely accurate. Time and again, experiment after experiment proved the power of the theory. Over the next few decades, quantum mechanics went on to gain nearly universal acceptance. Ironically, Einstein, though a winner of the Nobel Prize for helping discover the principle of the quanta, never quite totally believed in it. Recall his famous "God doesn't play dice" quote. While he couldn't deny the overwhelming evidence for the success of quantum mechanics, Einstein thought it to be an incomplete theory.

And it still is. After all, says Dr. Roger Penrose, it certainly works on one level, on the tiny, subatomic level, where distances are measured in millionths and billionths of a meter. But how does that translate to a world we can see and touch? "The problem is when you try to apply the rules of quantum mechanics for a large object." Erwin Schrödinger, the renown Austrian quantum physicist, pointed this out famously when in 1935 he created a thought problem, a paradox, dubbed Schrödinger's cat. "It wouldn't be hard, according to the rules of his own equation, to make a state in which the cat was alive and dead at the same time. And this, he says, is ridiculous." That's why, says Penrose, "quantum mechanics is extremely well confirmed and a beautiful theory, in many respects, and it certainly agrees with all observations that have been made so far. The only trouble with it is that it doesn't really make sense, and I think a theory ought to make sense."

So physicists were faced with a dilemma: They had two theories, two explanations, that were very successful in their own corners of the universe but could not describe the other's realm. Relativity theory—gravity—was good for explaining the giant scale of planets and stars and galaxies. Quantum mechanics was very good at describing the tiny world, but it could not explain the larger world we live in.

Hmm. What to do? Where to look for an answer? Would it help us reconcile the two if we could find a circumstance in which both ideas might be working at the same time? Of course it would. But when would that be?

How about the birth of the universe: the big bang, almost 14 billion years ago. Here, the universe and everything that ever existed would be infinitesimally small, so quantum mechanics would be in action. Yet things would be incredibly massive, so gravity would be in its glory too. All we needed to do, then, was to figure out what was going on at that time and the problem would be solved. Bingo!

And good luck.

Einstein spent the last 30 years of his life trying to unite the two ideas into a theory of everything. He failed. Ever since then, physicists have been trying to tread the same ground, trying to unite the two worlds. And so far, they have not succeeded. "I don't see much indication of it getting terribly simple just yet," says Penrose. "I think one likes to think that eventually there will be some simple principle that governs everything, but we're certainly a long way from that now."

And that brings us to today, where it appears that our universe continues to defy our understanding. We have the head-scratching discoveries about the composition of our universe—the mysterious dark energy and dark matter—and the complicated theories that try to explain these discoveries, the string theories and their extra dimensions.

So, as Rod Serling used to say, "presented for your consideration" are the challenges facing cosmologists as they try, like scientists for hundreds of years before them, to understand the world we live in. Cosmology is not easy to grasp in just one gulp. Over the years, I've had to read and re-read, listen and re-listen to scientists describe the spooky world of quantum physics just to make sure I heard and understood them correctly. That is why I've presented many views, from some of the world's leading cosmologists and thinkers. You'll note that they don't always agree, which makes cosmology one of the more interesting and fun topics of discussion.

CHAPTER SEVEN

IT'S A DARK WORLD AFTER ALL

Duct tape is like the force. It has a light side, a dark side, and it holds the universe together. . . .

—CARL ZWANZIG

Direct proof of dark matter: This composite image shows galaxy cluster 1E 0657-56, also known as the "bullet cluster." This cluster was formed after the collision of two large clusters of galaxies, the most energetic event known in the universe since the big bang. The relative movement of the two galaxies is the first direct evidence of dark matter. X-ray: NASA/CXC/ CfA/M. Markevitch et al.; optical: NASA/STScI; Magellan/University of Arizona/D. Clowe et al.; lensing map: NASA/STScI; ESO WFI; Magellan/University of Arizona/D. Clowe et al.

The 1990s have been called the golden age of cosmology as scientists gather new information about the size, the shape, and the makeup of our universe. Some of the research provides an update, such as the age of the universe now being set at about 14 billion years. Some of the research results are very surprising, such as the discovery that the universe is accelerating outward, propelled by some strange, repulsive dark energy that makes up two-thirds of the universe, or the continuing search for dark matter, which we can't see but we can feel. And in trying to explain these almost science fiction–like properties, researchers have had to turn their eyes from the very big to the very little, to the tiny world of subatomic particles, string theory, and beyond.

Dr. Lawrence Krauss, chairman of the physics department at Case Western Reserve University, says that while we have just finished a decade of "incredible revolution" in the field of cosmology, major questions remain to be answered. "We don't know what most of the energy in the universe is. We don't know what most of the matter in the universe is."

"When you look at the beautiful pictures on the Hubble Space Telescope, you're struck by the colorful galaxies and stars. What we know," says Dr. Michael Turner, chair of the astronomy and astrophysics department at the University of Chicago, "is that the visible part of the universe is just a tiny fraction of what's out there. Most of the stuff out there is dark. So if you think about it, the cosmic infrastructure is this dark matter, and it's just decorated by the pretty galaxies."

Dark? What about all of those stars and galaxies, the moon and planets that light up the night sky? Astronomers are saying that all the stuff we can see is only a small percentage of what makes up the universe. The rest of it is dark.

"We've had kind of a seventy-year-old detective story," says Turner, as he begins to tell his yarn. "It begins with Fritz Zwicky in the 1930s, who realized that the gravity of ordinary stars couldn't hold together clusters of galaxies," because there just wasn't enough of it. Zwicky, an astronomer born in Bulgaria who immigrated to the United States to work at California Institute of Technology, observed the universe through an array of telescopes and discovered in 1933 that there just wasn't enough visible mass in the universe to create the gravity to hold the galaxies together. But Zwicky was an eccentric guy and hard to work with, so scientists kind of ignored what he had to say, though they really couldn't punch holes in this observation.

Then along came Vera Rubin, in the 1970s. Rubin, an astronomer looking for something really interesting and unusual to work on that didn't conflict with raising her children, discovered that rotating galaxies did not behave the way they should. Matter at the outskirts of the galaxies should be moving more slowly than matter near the center, the same way that the planet Mercury, closest to the sun, rotates around it at a dizzying pace compared to the outer planets. The sun's gravity weakens as you get farther away, so planets slow down. But that doesn't happen in galaxies. Why not? There must be something—unseen or dark matter—whose gravitational force is making up the difference. But what could it be?

"In the 1980s, the theoreticians like myself and others," continues Turner, "honed down the story a little bit and said, 'Just a minute. There's not enough ordinary matter, the stuff of the periodic table, the stuff you and I are made of, to explain this dark matter, and so it must be something exotic." Something that is not made of atoms; something we have never seen, literally, before.

But even when they put on their best theoretical thinking caps, they could not theorize enough matter to make up for the missing gravity. It fell a full 75 percent short of what they needed! What to do? Well, if matter and energy are equivalent, why not theorize that what's really missing, or better yet, that what we can't see is not only dark matter but also some strange dark energy that had the property of being repulsive rather than attractive. In other words, dark energy was causing the universe to expand, not contract. If so, dark energy permeated the universe, everywhere, even here on Earth, right where you are sitting now, all around you. But theorizing dark energy and dark matter to make the equations of the universe work is one thing. Finding out, via observation and experimentation, just what they are is something else.

"Those are the biggies," says Steven Weinberg, a Nobel laureate in physics at the University of Texas at Austin.

These two problems have created heady times for physicists and astronomers, says Weinberg. "It certainly is a golden age for cosmology. It gives me a certain amount of pain to say that because I'm an elementary particle physicist, and we haven't had this kind of excitement in elementary particle physics for about, I'd say, twenty years at least, perhaps twenty-five years. Now we have a period when observations and theory are coming together, and experimentalists and theorists are talking to each other in meaningful ways, not just 'Where are we going out for dinner?' It's truly very exciting."

DARK ENERGY

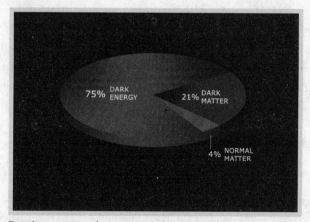

Data from measuring the temperature of remnants of the big bang reveal that the contents of the universe include 4 percent atoms, the building blocks of stars and planets. Dark matter comprises 21 percent of the universe. Seventy-five percent of the universe is composed of dark energy that acts as a sort of antigravity. Data were collected by the Wilkinson Microwave Anistropy Probe (WMAP).

Let's start with a real head-scratcher: dark energy. "Dark energy is even more puzzling than dark matter," says Weinberg, "because there are some reasonable candidates for what the dark matter should be. In fact, rather too many candidates. There are particles that appear in our theories that could have, if they existed, just the right properties to be the dark matter, whereas the dark energy is utterly mysterious." And contrary to what you might think, "the mystery is not why there is dark energy but why there's so little of it. In other words, it's not 'Why is the universe accelerating?' but 'Why is it accelerating so slowly?'

"You would expect, if you just did a back-of-the-envelope estimate, that the dark energy would be many, many orders of magnitude larger, and the mystery of why it is as small as it is, is really the bone in the throat of today's fundamental physics. No theory of physics has so far explained this remarkable fact. It's not mysterious that it should be there. In fact, any reasonable theory predicts it should be there, but it predicts vastly too much of it."

It's easier to eliminate what the dark energy is *not* rather than to surmise what dark energy *is*. Dark energy is not anything we know of today.

"Dark energy is not particles," says Weinberg. "That's the dark matter. The dark energy is something smooth. The dark energy is something with negative pressure. And that doesn't come out of any kind of particles. It comes out of a condition of space itself. It's an energy in space itself."

There's a concept that's hard to wrap your mind around: Empty space is not so empty after all. It has energy. And that energy has a force to it. Not an attractive force such as gravity, but a repulsive force that is causing the universe to expand. And if this repulsive force is not astounding enough, consider this bit of information: The dark energy did not show up until a few billion years ago. "The amazing thing is that it wasn't there from the beginning." If it had been, says Weinberg, "the universe would have expanded too rapidly to ever form galaxies. Which means no stars, or planets, or life. We wouldn't have been here."

And if that idea is not enough to make your hair hurt, here's the kicker: *The repulsive energy did not reveal itself until a few billion years ago.* Yes, even though the universe is about 14 billion years old, the dark energy did not show up until the universe was about 8 billion years old.

"On the theoretical side," says Turner, "we're very, very excited because this is a big problem, and theorists like big problems. And big problems require new ideas, ideas that at the time may seem crazy. And so both sides, both the astronomers telling us about the acceleration and the theorists making up ideas, are having a great time. What's really exciting is that it's a game that we can all play. We don't have a clue as to what's causing the universe to accelerate, but we know it's extremely important and may provide a key to understanding how the forces and particles of nature are unified."

THE SEARCH FOR MORE DIMENSIONS

"Some of the solutions that have been suggested involve the fact that maybe the universe has more than four dimensions, and maybe out there, in those other dimensions, things are happening that are influencing the four dimensions that we're privy to, and that maybe the speeding up of the expansion has something to do with the existence of other dimensions," says Dr. Lisa Randall.

We're all comfortable with the idea that the world we live in has at most four dimensions: three spatial and one time. Randall has been thinking about a universe with as many as nine *extra* dimensions, hidden dimensions that we can't perceive. She is a professor of theoretical physics at Harvard University and the author of *Warped Passages: Unraveling the Mysteries of the Universe's Hidden Dimensions.*

"One of the things that makes it difficult to get it across is that you can't picture it. We are not physiologically designed to picture more than three dimensions. It doesn't mean they're not there, but we certainly can't just picture them very simply. Really, the right way to understand it is with words or equations. It's hard to understand it with pictures, because we just can't see them."

So the process of thinking about extra dimensions becomes a mind game, a "thought problem," as Einstein used to call it.

"Sometimes I'll just sit on my sofa and think. Sometimes I'll just talk to other people and exchange ideas. A lot of the time we'll have the blackboard filled up with equations or pictures, where we try to have some back-and-forth about ideas, or sit in a coffee shop. Sometimes I will just sit down and work out the equations, sometimes you'll sort of talk to the equations. You'll try to figure out, what are they trying to tell you? Or you might have some idea of where you're going when you use your equations, so where is this all heading?"

The search for evidence of these dimensions has united the worlds of the very big and the very small, those seeking to understand the universe on a cosmological scale and those seeking to reveal the unseeable, subatomic world of particle physics. Both worlds are looking for evidence of these extra dimensions to satisfy their own hunger to explain how the universe was formed and where it is going.

"We have quantum mechanics, which describes things with small scales, atomic scales," says Dr. Randall. "We have general relativity, the theory of gravity, that describes things on big distance scales. And these theories work fine, but at some infinitesimal distance, not a distance we're going to experience—ten to the minus thirty-three centimeters—the theories are incompatible. That tells us that there's something wrong with the theory, and we'd really like to have a theory that can describe everything."

Ah, yes. The quest for a theory of everything. A theory that would unite all the forces and matter in nature. One of the great problems still challenging physics is trying to unite gravity, which Einstein described as curved space and works over long distances, as Dr. Randall says, with the other forces of nature, which are described as quantum particles—subatomic particles such as quarks, bosons, leptons, and neutrinos. The key to answering this problem is understanding why gravity is so weak compared with the other forces in nature.

"It is kind of remarkable that you can pick up a paper clip with a magnet when the entire Earth is pulling against it. And from the point of view of an elementary particle, gravity is just completely negligible, compared to the other forces. In fact, it's very hard to test gravity because the other forces swamp it so much. And it's even worse than that because in order to unite gravity and quantum physics, we actually have to introduce such a big fudge in this area that we know there's something else. And that's why this question is really driving a lot of research today: How do you unite the two?"

Can you find particles that describe gravity? The answer to that problem, so far, has eluded just about everyone, from Einstein on. Scientists have been looking for this "theory of everything" for quite a while. "We haven't yet worked it out," quips Randall. "We don't even really know, in many ways, what the theory is."

STRING THEORY

Randall is referring to one of the most widely talked about theories of everything: string theory. It's the idea that instead of tiny "particles" being the building blocks of everything we see and feel, the universe can be explained by tiny, unseeable "strings" that when "plucked," vibrate in so many different ways that they create the forces and stuff of nature, even solve the gravity–quantum problem. We'll explain this in greater detail in the next section. But for now, suffice it to say that string theory requires the existence of all those added dimensions. Perhaps 11 of them or more. And yet even with those extra dimensions, says Randall, string theory does not tie up the loose ends of a theory of everything.

"In fact, one thing we've found is it's not even just a theory of strings; it has other exotic objects called *branes*, which are membranelike objects. So the theory clearly has a richness to it, but we don't know exactly where it's leading us."

"I think it's important for people to realize that not only has this idea [extra dimensions] cropped up many times," says Lawrence Krauss, author of *Hiding in the Mirror: The Mysterious Allure of Extra*

Dimensions, from Plato to String Theory and Beyond, "but that in some sense, we're kind of hard-wired to want there to be more out there than we can see."

Krauss is not convinced that string theory will be successful or even necessary to solve problems in physics. Or even that extra dimensions may be necessary to explain our existence.

"The world is a terrifying place, and I think when humans first evolved, it was clearly terrifying, and the hope that there was some better place where things might be fairer was certainly a part of it. And this is a time when science is again looking at this intensively, and serious people are saying there may be ten, eleven, twenty-six dimensions. But maybe the fact that it keeps cropping up even in science has more to do with our psyche than with the universe."

Randall disagrees, "We know ideas like extra dimensions and these things called branes (strings that are stretched long and flat) might actually be a part of our universe, and so you can't just decree that we're not going to think about it. On the other hand, we *do* want to make connections to the world. So really what we want are two simultaneous directions. We know this theoretical problem of trying to reconcile gravity and quantum mechanics is there, but we also know that there are phenomena that we don't understand, even at observable scales, which is why we want to build accelerators like the Large Hadron Collider [LHC] that will be at CERN [European Organization for Nuclear Research] in Geneva."

Randall is optimistic that experiments, like those to be conducted at the LHC in 2008, may answer some of the unknowns. The LHC is the world's biggest atom smasher, or as they say today, particle accelerator, built near Geneva, Switzerland. In it, two beams of subatomic protons will race in opposite directions, underground, in a 27-kilometer-circumference circular tunnel. Gaining enough energy, they will smash into each other at tremendous speeds, in the hopes of producing even more subatomic particles that may answer some of the riddles of nature, such as "Can we find evidence of these extra dimensions?"

Large Hadron Collider: Computer simulation of an experiment at CERN showing the detection of the Higgs boson, a particle that—theoretically—imbues matter with mass.

Experiments at CERN will look for evidence of "supersymmetry." Supersymmetry is a very popular idea suggested by the unification of the forces of nature. It says that for every particle there is a supersymmetry partner. If tests at CERN find evidence of these supersymmetry partners, then it adds weight to the validity of string theory.

"It doesn't mean we'll directly test string theory," says Randall, "but perhaps we can test ideas that come out of string theory. I think it will give us ideas of where to head with string theory. If we discover double the number of particles, that would tell us something about that symmetry holding to very low energies. If we discover extra dimensions, it would tell us something about what those dimensions look like. So I doubt very much it will tell us whether string theory is right, but I think it could tell us things that might guide string theory research in the future."

NOT YOUR FATHER'S TELESCOPE

While the particle physicists are keeping their fingers crossed on finding evidence of supersymmetry at CERN, cosmologists are betting that they can beat the particle physicists at their own game by using their telescopes to discover evidence of supersymmetry in nature's own particle colliders: the black holes, energetic hot stars, galaxies, and supernovae of deep space. Even the echoes of the big bang that still reverberate around the universe—the dark microwave background radiation—offers clues.

"One thing we would all probably agree on is that there are deep connections between the elementary particles and the universe," says Turner. "When we talk about the birth of the universe, we need ideas from the elementary particle physicist, and then we need to turn those ideas around and see how they can be tested with the sky today. And when we look at the solutions to the big questions involving dark matter, we hope to produce [it] at an accelerator. So it's not just telescopes. It's also accelerators."

Unlike theorists who hang out by their equation-filled blackboards and think heavy thoughts, the experimentalists actually do the heavy lifting. They must search for evidence in the cosmos that would back up the repulsive nature that the theorists are predicting. "The observers now have the ability to probe the acceleration of the universe," says Turner.

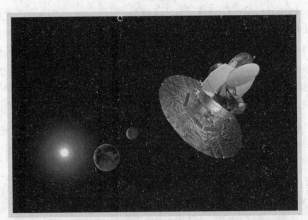

WMAP (Wilkinson Microwave Anisotropy Probe) spacecraft, which measured the temperature of the universe. NASA/WMAP Science Team.

Some of the tools astronomers are using are the "way back" machines of astronomy: telescopes. In practical usage, a telescope is really a time machine, says Turner. "It allows you to look out in the universe, and as you look out, you look back in time. And so you can compare what the expansion rate is today with what it was back then."

But the telescopes they use are not the kind you use in your backyard to look at the moon. "Some of them will use X-ray telescopes." Some will use telescopes that can see gravity acting as a lens to bend light around galaxies. "Some of them are going to use the microwave background to get at this question of acceleration."

The microwave background radiation is the "echo" left over from the big bang. A NASA satellite, the WMAP (Wilkinson Microwave Anisotropy Probe), hovering in space between the Earth and the sun, has been able to peer back in time to almost the beginning of our universe, detect this very faint microwave remnant, and produce a detailed picture of its early history. "It amazes me that we can say anything about what transpired within the first trillionth of a second of the universe, but we can," said Charles L. Bennett, WMAP principal investigator and a professor in the Henry A. Rowland Department of Physics and Astronomy at Johns Hopkins University. "We have never before been able to understand the infant universe with such precision."

So far, in a world where no one knows just what the dark matter or energy is, Dr. Neta Bahcall, professor of astrophysics at Princeton University, says the observations are "remarkable," because they are all consistent. "All the different observations from different methods—from weighing the universe to measuring distant supernova, to the microwave background radiation—all of those very different methods yield the same result. They all show that we made dark matter in the same amount as we expected. They all are consistent with having some amount of dark energy,

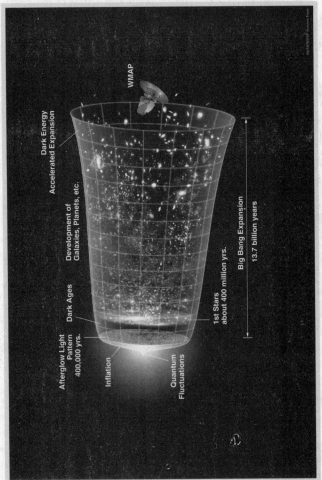

A timeline of the universe. Courtesy of the NASA/WMAP Science Team.

and that's why it is so exciting. If you talk to the astronomers, we are just jumping for joy, and really jumping in our seats because it is a very exciting time to see that the data is all getting together so beautifully. What is a dark matter? What is a dark energy? What does it all mean for physics? We're going to have answers in the next ten years or so."

DARK MATTER: AXIONS, WIMPS, AND MACHOS?

Some of those answers may come not from looking up into the heavens but from looking down into the basements of giant particle accelerators. In their search for the solutions to the super-big questions of the universe, physicists are probing the super-tiny world of the atom, looking for the particles, about whose existence they can only speculate.

Among the most popular candidates are *axions*—cold particles predicted to have been created in abundance during the big bang. Axions were named after a laundry product by a copredictor of the particle's existence, Dr. Frank Wilczek. He believed that their presence would help "clean up" some of the problems in theoretical physics. *Wimps* are another creative solution to the dark energy problem. The term is short for "weakly interacting massive particles." Wimps theoretically cannot be seen directly because they do not interact electromagnetically or with atomic nuclei. But because they are cold, massive particles that would tend to clump together, wimps fit the bill for cold dark matter. All we have to do is prove they exist.

Finally, of course, if you have wimps, then you need their alter ego: *machos*, "massive compact halo objects." (Those physicists are such cards!) A macho would be a clump of normal atomic matter, such as protons or neutrons, that float unseen through the halos around galaxies. They might be black holes, neutron stars, brown dwarfs . . . yada yada yada. You get the idea; the possibilities are many.

"With all of the instruments we have, two satellites to map the echo of the big bang, the Hubble Space Telescope, the Chandra X-ray Observatory, and NASA's next-generation space telescope [the James Webb Space Telescope] that will have ten times the light-collecting power," Turner thinks that cosmologists will have the riddle of the dark matter and energy solved within a decade. "Some of them will use X-ray telescopes. Some of them will use telescopes in a very intriguing way where they look at how the distribution of matter between us and the distant universe is bent or gravitationally lensed. Some of them are going to use the microwave background to get at this question of acceleration. I can't predict the exact date, but in the next ten years we're actually going to be able to answer these difficult questions."

"As we use this high technology to zero in on every little detail of the matter in the universe, we'll see that infrastructure in stunning detail, and that will reflect the type of dark matter," says Dr. Andreas Albrecht, professor of physics at the University of California, Davis.

"An important point to make is that the problem with dark matter is not a particle physics problem," says Dr. Bernard Sadoulet, professor of physics and a director of the Center for Particle Astrophysics at the University of California, Berkeley. "It is an astrophysical problem, which may point to questions of deep solutions for the particle physics problem. But it's really the merging of these two fields, of cosmology and particle physics."

"Discovering supersymmetry has been the holy grail for a number of years of the particle physicists," notes Turner, "and it looks like the cosmologists—I've got my fingers crossed here—might beat them to it, because it might be that our galaxy is held together by supersymmetric particles, and so I'm putting my money on people like Bernard, that they're actually going to beat the LHC to finding supersymmetry. But as Sadoulet says, in the end, both the astrophysical evidence and the evidence in the laboratory will make it a much richer picture."

This is a remarkable idea: that the largest things we can imagine—galaxies stretching millions of light-years—started out as small, subatomic particles. "And even more remarkable than that," says Turner, is that "we can test it. And the answer seems to be written across the microwave sky that the biggest things did evolve from the smallest things."

CHAPTER EIGHT

STRING THEORY: WE HAVE A PROBLEM

String theorists don't make predictions, they make excuses.
—RICHARD FEYNMAN

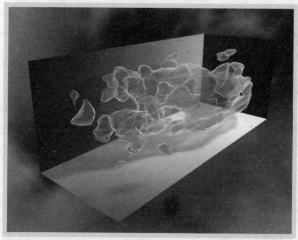

A three-dimensional map of distribution of dark matter in the universe, created by taking slices through different regions of space. The map stretches halfway back to the beginning of the universe. NASA, ESA, and R. Massey (California Institute of Technology).

Sometimes you hear something so often that you think it *must* be true. You wish it were true because it would solve a lot of problems (does "weapons of mass destruction" sound familiar?). That's not supposed to happen in science. Ideas and theories are put to the test, to find out if they can stand up to scrutiny. If they do, they become an accepted part of the way we look at the world, until something better comes along. Relativity theory is one such idea. It expanded on a really nifty theory of gravity put out by a guy named Newton, hundreds of years before it. Reality itself has been under scrutiny and test for more than a hundred years, since a pretty good physicist named Einstein published the first of his many papers about relativity in 1905.

Isaac Newton was the rock star of science in his time. And relativity theory—more accurately, Albert Einstein—was the darling of the media immediately after a famous eclipse of 1919 proved true his revolutionary theory that gravity curved space. The modern equivalent would be string theory. Countless books, papers, articles, and TV shows have sought to explain it to us; it has made media stars out of scientists, such as Brian Greene, professor of physics and mathematics at Columbia University in New York. He's also the author of several books, including *The Elegant Universe* and *The Fabric of the Cosmos*.

STRING THEORY IN A NUTSHELL

According to Greene, "We have two main pillars of understanding in physics that were developed in the twentieth century. One is the general theory of relativity, Einstein's theory that describes gravity. And gravity is a force that's relevant mostly when things are big—stars and galaxies and so forth.

"The other major development is quantum mechanics, a theory that describes the other end of the spectrum, the small things—the molecules and the atoms and so forth.

"Now for a long time, we've recognized that these two theories have to talk to each other in a sensible way. There are realms, extreme realms, where things are both heavy and small, like black holes or the beginning of the universe. And because those realms exist, you need to use gravity, general relativity, and quantum mechanics all at the same time.

"The problem is that for many decades, any attempt to put the two theories together, to unify them, didn't work. It gave wrong, nonsensical answers. String theory is an attempt to fix that, to give us a theory that won't give nonsensical answers, that will give answers that make sense when you put gravity and quantum mechanics together."

QUANTUM RELATIVITY

In a nutshell, string theory says that instead of tiny little subatomic particles being at the root of all matter and energy, tiny little strings exist instead. And when plucked just the right way, they vibrate to produce the building blocks for everything we see around us. And most importantly for cosmology, string theory, theoretically speaking, can unite the disparate worlds of gravity and quantum mechanics.

Sounds great so far, doesn't it? But as with anything else in life, string theory comes with a little bit of extra baggage. A few big problems. Because while string theory may look sweet, mathematically speaking, it requires that our world be composed not of the 3 or 4 dimensions familiar to all of us but instead 10 or 11 dimensions, most of which remain hidden from view. "Whether there really are extra dimensions or not, I think, remains to be seen," says Dr. Lawrence Krauss.

Even Einstein found no way to unite gravity and quantum mechanics. And that's where some scientists see the problem: String theory is aging, and not elegantly. It has been around long enough to be tested, and we should have found experimental evidence to prove that it works. And because it hasn't been tested, scientists, in a rapid sequence of speeches, articles, and books, are beginning to openly question its usefulness in physics. They say that it is not living up to its promise, that it's more hype than science.

Take Lee Smolin, faculty member at the Perimeter Institute for Theoretical Physics in Waterloo, Ontario, Canada. Smolin, of *The Trouble with Physics: The Rise of String Theory, the Fall of a Science, and What Comes Next*, says he is trying to understand why string theory, which he has worked on himself, is so problematic. "Why the ideas which seemed at first so beautiful, so natural, were not getting us where we expected to get twenty years ago."

Smolin points out that just about every 25 to 30 years, new ideas in physics come along to replace the old ones. If you look back over the last 200, he says, it's very unusual for three decades to pass without a scientific revolution. For example, he writes that "from 1830–1855 Michael Faraday introduced the notion that forces are conveyed by fields, an idea he used to greatly further our understanding of electricity and magnetism."

In the 25 years that followed, James Clerk Maxwell expanded those ideas into "our modern theory of electromagnetism." He explained how light was also like radio waves and unlocked other secrets of our natural world. Then came the next dramatic period, 1880 to 1905, in which electrons and X-rays were discovered. In that 25-year period, Smolin continues, Max Planck's work would "spark the quantum revolution." Einstein's era would arrive in 1905, and understanding the impact of relativity would occupy the next two and a half decades. By Einstein's death in 1955, we would know all about a whole new world of subatomic particles and organize the forces of nature into a family of four.

The next 25 years and the 25 after that would see the creation of a "standard model" of the elementary particles in the universe, and on a larger scale, we'd see Stephen Hawking and other luminaries enlighten our understanding of black holes, the big bang, and dark energy and matter.

But since the 1980s, says Smolin, we have been stymied. String theory is now more than 20 years old and doesn't seem to be yielding the answers that have been expected of it. "History seems to show that when there's a good idea about unifying different parts of physics, it works fast if it's going to work." Scientists should be able to conduct experiments that either bolster or bat down that new idea. But the problem with string theory is that so far, there are no experiments that can solidify it. Or, as Smolin puts it, "string theory is not making experimental predictions. There are certainly very beautiful things about it," but a theory that can't be tested is nothing more than a theory. In science, you can't hang on to an idea too long; if it can't be tested and proven, it's on to the next big thing.

Krauss agrees. He says string theory has been a failure. "There isn't a shred of empirical evidence, not only for extra dimensions but essentially also for string theory. They [scientists] haven't made any predictions that have been tested. And moreover, in fact, to some extent, we're still just learning what the theories are."

Greene agrees that the evidence, so far, has been lacking. But as one of string theory's greatest proponents, he says we have to give the experimentalists a bit more time to find the evidence.

"How long will it take for those experiments to happen? I don't know. We could get lucky. It could be that the Large Hadron Collider, which will turn on in 2007 or 2008," will show some evidence for string theory. "We might see some of the fingerprints of string theories through something called supersymmetry, certain particles that the theories suggest should be there but nobody has yet seen."

Astronomers, looking into space, might also find evidence of string theory. "That's what I spend my time on these days, trying to see where these strings might leave some imprint in the microwave background radiation, the heat left over from the big bang. All of these are long shots. But we're doing exactly what Lee is saying one should do in science—namely, work toward experimental verification. How long? I can't predict. Nobody can predict."

Greene says it could be years, even decades, before experiments can determine the validity of string theory. And should that happen, interest in the theory would drop off "because the people who work on string theory are physicists, and physicists want to make contact with physical reality." But what has happened since the 1980s is that the theory has gone through "what we call revolutions in our thinking time and time again, which has given a surge of energy, a surge of interest in the theory, which has kept us going even though we've yet to make that desired contact with experiment. Our understanding of the underlying theory, our understanding of the equations, our understanding of the fundamental ideas and how they relate to one another—we've made great strides.

"We have an international meeting every year, a string theory conference that has something like fifty talks. And these talks are generally amazing. They're generally showing how people are making great progress in spite of not having the guide of experiment. So if it turns out, as Lee is saying, that some of the experiments he's describing or the experiments of the Large Hadron Collider, if in the next couple of years these experiments bear fruit and begin to show us some of the features of the theories that we've been working on for a long time, things will definitely take a major leap forward.

"So it's a very exciting time, waiting to see the results of those experiments. And in no way would one want to say that the theory is moving slowly. It perhaps is moving slowly toward these experiments, which are coming online. But the theory itself is developing rapidly. In fact, it's hard to keep up. The theory is able to embrace all of the major developments in physics having to do with the elementary particles in quantum mechanics that were discovered before string theory in the middle of the twentieth century, leading up to the end of the twentieth century. They all naturally find a home within string theory. And that's very compelling to us because usually a revolution doesn't actually erase the past. It embraces the past but goes further. And that's what string theory seems to be doing.

"The other side of it is that even without the experimental confirmation, string theory has a very intricate mathematical structure that holds together with a kind of tight, logical cohesion. There are checks and rechecks in the calculations, enormous number of consistency checks, and they've all passed. The theory comes through with flying colors every step of the way. And that again keeps us going, keeps us thinking that this theory is at least heading in the right direction."

To hear Krauss talk about string theory, it sounds more like a solution in search of a problem.

"It still amazes me, when you think about it, that string theory arose in the 1970s when people were trying to understand all this host of new elementary particles that were being discovered in accelerators and they couldn't make sense of it. This theory came along that looked like it might help you make sense of it, but, by the way, it required twenty-two extra dimensions. And I'm amazed in some sense that physicists were willing to automatically assume that maybe all those extra dimensions exist just to solve that problem. It turned out it wasn't the solution to that problem. But then a decade later, physicists realized maybe it was the solution of another problem involving gravity. And physicists, many of them, are convinced those extra dimensions are out there.

"And to the credit of the physics community, there are some people who are actually trying to think of experiments that might actually be able to test this, so it isn't just metaphysics."

IS NEW SCIENCE SUFFERING?

Smolin says that he would never tell Greene or anyone else working on string theory to drop what they are doing and head into something else. "Certainly time will be the judge. If somebody feels that string theory or anything else is the most promising thing they know about, certainly they should work on it.

"But there is another level, and that's the level where we think about science as a very risky activity. And if it is a very risky activity, something like development of a new technology, the question arises: Do we support only one direction? Do we put all of our apples or whatever it is in one basket? Or do we hedge our bets? Do we support all the people who are excited about the good ideas that they have?"

And this is in large part an issue that concerns Smolin. "It's not a question of what one individual scientist does. One individual scientist should do what he or she deeply believes in." But it is a question of science as a community, where so many scientists are working only in one direction: string theory. That kind of community does not encourage scientists to strike out in other directions, on their own, where historically new ideas arise. "The analogs of the great physicists of the past who always struck out on their own, people like Galileo and Einstein, those people didn't have an easy time because of the way that universities are very averse to risks. They're very averse to hiring people who are working on their own ideas as opposed to ideas that large communities of people have been working on for decades. We should try to find ways to help and support those people who have new ideas and have the courage to work on their own ideas."

As a community, "we can take attitudes where we encourage people to strike out on their own, to leave behind old ideas, even if [there are] still things about them we love, and to encourage the young people, especially the young people, to forget what people of our older generations have done and strike out for new directions."

On this point, there is no disagreement. "On this other issue of encouraging young students to strike out on their own and pursue their own ideas, I couldn't agree more," says Greene. "I, for instance, in the last couple of years have had students that don't work on string theory. I've had students that have worked on relatively fringe ideas, according to the mainstream point of view. I've had students working on more bread-and-butter particle physics. So absolutely, we need to encourage diversity of thought. We need to encourage the young students to express their creativity. Who would ever say otherwise?"

THINKING OUTSIDE THE BOX

And just what kinds of "diversity of thought" might one find? What is occupying some of the best young minds? Smolin points to a few mind-numbing approaches:

- *Deformed special relativity:* It's the idea "that quantum gravity alters the basic equations of special relativity in ways that are testable by experiments" to be performed in the near future.

- *Dynamical triangulation:* "One of several ideas on the basis of which space is made of discrete elements. One tries to find effects that come from the hypothesis that space is discrete," as opposed to being the continuous, smooth place we experience it to be.

- *Loop quantum gravity:* Smolin's own area of research, "something that is a successful unification, at least at the level of the equations, of general relativity and quantum theory. It has led to a very particular picture of space being made out of discrete elements. And there are consequences of that which people are exploring." This theory says that the big bang was *not* the beginning of time, so that time continues into the past.

What should we expect once we can unite quantum mechanics with Einstein's concept of space? Some very interesting results because of the difference in the ways the two act. Quanta like to "leap" in discrete jumps, and quantum particles can appear in many places at the same time—even tunnel through things. But our concept of space is one of smoothness. Objects travel through space in smooth lines, sailing on a continuous, uninterrupted trajectory from the Earth to the moon, for example.

So uniting the two worlds would lead to some ideas that seem to come right out of science fiction. You get sort of a hybrid of the two. "The notion of space should disappear," says Smolin, "just like the notion that the trajectory of a particle disappears in quantum mechanics." Instead, you get the spooky world of a quantum state, where "a particle is either a wave or a particle, depending on what questions we ask about it. The idea that we're living in this three-dimensional, fixed geometry goes away."

REVOLUTION IN PHYSICS

Where is all this leading to? Smolin is unequivocal. "I think we do need a new physics. We need to complete a revolution. Einstein started this. Einstein started the revolution in the early 1900s when he was the first person to declare that we needed a quantum theory to break with the physics that went before. And he also brought us relativity theory. And that was the launch of the revolution. And we're still engaged in that same revolution. It won't be over until this problem of putting together relativity and quantum theory is solved, and not just solved in principle on a pad of paper but solved in such a way that it leads to new experiments and new predictions for experiments."

Greene goes even further. "I full well believe that when we do complete this revolution that Lee's referring to, we will have a completely different view of the universe."

But there's more. "I totally agree with Lee that everything that we know points to space and time not even being fundamental entities." Now that *is* revolutionary. Space and time no longer the basic building blocks, yet we think they are? Greene shows why he is such a good explainer.

"The way I like to think about it is to take the concept of temperature. We all know what it means for something to be hot or to be cold. We can experience it. But scientists taught us that there's an underlying physics to temperature which has to do with how fast particles, molecules, are moving. Molecules move fast, it appears hot. It feels hot. Molecules move slowly, it will feel cold. So the idea of temperature rests on a foundation of more fundamental ideas, motion of molecules.

"We think that space and time are like temperature in the sense that they rely upon more fundamental ideas as well. Now what those more fundamental entities are—the so-called atoms, if you will, that make up space and time—we don't know yet. String theory has some vague suggestions. Loop quantum gravity has some vague suggestions. We're not there yet.

"But when we get there, I think we will learn that space and time are not what we thought they are. They are going to morph into something completely unfamiliar. And we'll find that in certain circumstances space and time appear the way we humans interpret those concepts, but fundamentally the universe is not built out of these familiar notions of space and time that we experience."

Not only would it shift our view of physics but it would also change the way we look at the world.

"It would change the very notion of reality, if you really want to be more precise, because most of us, at least, think about reality as existing in a region of space and taking place through some duration of time. If we learned that those basic ideas, the arena of space and the duration of time, are not concepts that even apply in certain realms, the realm of, say, the very extreme of energy or the very extremes of small size and tiny intervals . . . if the notions of space and time evaporate, then our whole conception of reality, the whole container of reality, will have evaporated and we'll have to learn to think about physics in the universe completely differently."

To my way of thinking, if there were ever a case of fact being stranger than fiction, this would be it. But don't get carried away, says Smolin.

"It's not—it's not as far out as it sounds. Really, it's not. And seeing space is made up of something—'atoms' of something more fundamental—you can go a long way with an image that the air looks smooth, the water looks smooth, and we discover that really it's made out of atoms."

And one great consequence of that image, says Smolin, is that it allows us to experiment with the concept, to see if it stands up to scrutiny. We can search for evidence of those "atoms" of space. "And that's why I keep pushing about experiment so much, because indeed it does seem that if space is made of atoms, there are consequences for how light propagates. And these consequences are checkable by experiments that use observations of light coming from very, very far away to look for very small differences in how light of different colors or different energies propagate."

It doesn't matter, he says, that we don't have a theory that predicts "an absolutely precise prediction for what these experiments should see. And that's a source of great frustration to all of us. But we have some general ideas about what the effect should be to look for, and if those effects are seen—and this could be in as little as a year and a half, two years, from this experiment that I was talking about that looks at gamma rays coming from very, very far away—then that will indicate that what Greene is saying is absolutely right. That will be the discovery of the atoms of space in the same way that some of Einstein's discoveries really cemented the idea that matter is made of atoms."

PART III

—

GETTING READY FOR
GLOBAL WARMING

CHAPTER NINE

IS EVERY COASTAL CITY A NEW ORLEANS
WAITING TO HAPPEN?

*The human race is conducting a profound and largely irrevers-
ible experiment on world climate. Increasing atmospheric con-
centrations of carbon dioxide and other greenhouse gases are
expected to cause a global warming that could raise the sea several feet in the next century. Should we wait for this experiment
to unfold, or prepare now for its possible consequences?*
—WILLIAM D. RUCKELSHAUS

The problem with speculating about the future is that tomorrow will
be yesterday by the time you read this. Words fixed on paper cannot
keep up with the dynamics of a world that is changing by the minute. The future may change overnight to become the past. Perhaps
next time it will be not the Gulf Coast under hurricane attack but
the Florida Keys. Or then again, maybe Mississippi, or even, yes,
New Orleans once more.

But it really doesn't matter because sooner or later, just about all
coastal towns around the world will be flooded, victims of melting
polar ice and rising sea levels, the signature of global warming. The
vast ice sheets of Greenland and Antarctica are melting, some of
them at twice the rate expected. Just the runoff of the melting ice
from Greenland, flowing into the North Atlantic, could raise sea

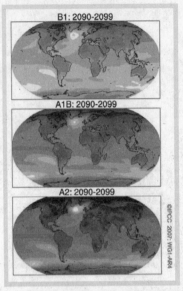

Projected surface temperatures for the next hundred years, 2090–2099, relative to the period 1980–1999: Three scenarios reflect differences in the data collected. Darker areas represent rates of global warming. All three predict much higher temperatures in the next century. Intergovernmental Panel on Climate Change, 2007.

levels 20 feet in the next few hundred years. And it won't be just the coastal cities of America sinking beneath the waves.

"If you want to see what Shanghai, Gdansk, Poland, Bombay, India, and New York City are all going to be obsessively dealing with fifty, seventy-five, a hundred years from now, turn on your television and look at New Orleans, says Mike Tidwell, activist and environmental writer (*Bayou Farewell: The Rich Life and Tragic Death of Louisiana's Cajun Coast*), "because that same sea-level rise from global warming will wipe out barrier islands, destroy buffering wetlands, and cause all kinds of problems to cities that are designed and built along the East Coast.

"Because of global climate change in this century, we will see between one and three feet of sea-level rise worldwide. Whether the land sinks three feet per century, as in New Orleans, or the sea level rises three feet per century, as in the rest of the world, you have the same problem."

So what are the low-lying seacoasts and ports around the world going to do about it? How will they survive? It appears that they have a couple of options: They can try to do what the Netherlands has done and build dikes and seawalls and hold the sea back. Or they can retreat inland and allow the sea to capture—or recapture—real estate destined to be underwater in a world of rising sea levels. About one quarter of the Netherlands is below sea level, yet it has engineered a system of levees and seawalls that keep the sea under control. Why can't we emulate the Dutch?

IMITATING THE DUTCH

"I'm really encouraging everybody to look at the Netherlands," says Ivor Van Heerden, director of the Center for the Study of Public Health Impacts of Hurricanes and deputy director of the LSU Hurricane Center at Louisiana State University in Baton Rouge. "When you compare our levees to theirs, it's almost shameful. They've got it

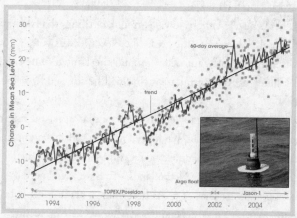

Oceans are getting warmer: This chart reflects the upward change in ocean temperatures over the years. Instruments such as floating buoys (insert) measure water temperature.

right. They've got lots of technique. They've combined hard structures and earthen levees, compartmentalization. They've got some excellent pump technology. We could learn a lot from them.

"But the 'plus' that we have in Louisiana that the Netherlands doesn't have is we have our coastal wetlands—what's left of them. We have barrier islands, and we have the sediment of the Mississippi River. So combining the levees—looking at the Dutch, learning from them, combining the levees with wetlands, with barrier islands, gives us, in essence, a three-tier level of protection. The barrier islands protect the wetlands. The wetlands protect the levees, and the levees protect the homes and infrastructure. And I really hope that we look at the Netherlands and use some of their technology as we move forward."

Dr. Robert Bea, a member of the Independent Levee Investigation Team, and a professor of civil and environmental engineering at the University of California, Berkeley agrees. "The water is the enemy and we need to find out how to slow it down, number one; and number two, when it starts to get out of hand, we need to be able to control the water levels and the wave action. So gates in some places are things we want to think about, in addition to levees, pumps, and evacuation techniques.

"There's a wide variety, a wide arsenal, of things that we have at our disposal to help not only areas like New Orleans but also other areas that have important deltas around the United States and, of course, many of the coastal areas that will be stressed by rising sea levels.

"In fact, I gave this problem to my class that works on risk management," recalls Dr. Bea. "We looked at New Orleans and came to the conclusion that based on economics and standards of practice and historic precedent like the Netherlands—that we would need to go for a ten-thousand-year level of protection in most areas. It could be one thousand in other areas, depending on what we were protecting."

Bea and other experts agree that the standards set for New Orleans are too low; they need to be raised to withstand the kind of ferocious storm that comes not once every 100 years but once every 10,000, the same standards—again—that the levees and gates in the Netherlands are held to. He challenged his students to find a way of meeting that standard.

"The next step we took was: How do we do that? And at this point, we started thinking about big levees, levee heights of forty or fifty feet started to show up. And at this point, you say, 'Hey, this isn't sustainable.' If you try and pile up something like this in a permanent works, with subsidence and settlement, you can expect it to go down. So at that point, we had to abandon the traditional—what I call brute-force—approach and start thinking about defenses in depth."

REVIVING THE WETLANDS

Defense in depth, to scientists, means that you have to bring nature back into the plan. Concrete and steel barriers will just not be enough to hold back rising waters, whipped by hurricane-force winds. Bea says the first line of defense would be strong barrier beaches to first slow down the inward rush of storm surge.

Then, "let's use those wetlands to help absorb some of the turbulence and energy out of the water. That's been called horizontal levees. And then, by the time we get to the perimeters we're trying to defend—this third level—we've now got earthen levees, and in some cases, indeed, gates, where we can stop the water from getting into areas around New Orleans like Pontchartrain and Lake Bourne. And at that point, you say, 'Well, even then I've got to have pump stations to get water out of these low-lying areas.'

"What you're doing is you're developing a system, and the system is one that starts with nature—and so uses the natural defenses and then complements nature with engineered works, so that we have a system in place that is environmentally friendly and—very, very, importantly—we can afford it and sustain it."

Not everyone agrees that the Dutch have all the answers. Not Mike Tidwell.

Georgia coast showing flooding that would happen if global warming were to raise sea levels: Black areas show flooding at high tide for a rise of two feet in next 100 years. Tidal variations and sinking of land are also factors. J. G. Titus and C. Richman, "Maps of Lands Vulnerable to Sea Level Rise: Modelled Elevations along the United States Atlantic and Gulf Coasts."

"A lot of people look at Holland and say, 'Well, they've got it going on. They've reclaimed the sea and they've built these seawalls and they've mastered the natural forces of this coastal area.' And the reality is that that is just as great an environmental calamity and a social infrastructure waiting to implode just like Louisiana, because the land in Holland is sinking just as rapidly as in New Orleans because they've done the same thing. They've leveed their major rivers along their coasts, and there is no land building going on. And even though they're behind those seawalls, it's not a sustainable way to go forward in the future. It's just not something that you can count on a hundred years from now. Sooner or later, sea-level rise and subsidence of that land will make that whole living arrangement in Holland unsustainable, believe me."

PLANNING FOR THE NEXT CENTURY NOW

In fact, not only the Netherlands is sinking but so is New Orleans. So in many cases, nations facing global warming and rising oceans are struggling just to stay in place, to make up for sinking real estate while planning for sea-level rise. It's a problem that can only get worse by the minute but will take decades to solve, making now a good time for cities to start taking action for the future.

"That is exactly right," says Bea. "We need to make prevention number one on our list. It's going to take a long time to get ahead of the problem in a sensible way. I think the other thing that we need to recognize is when to give up. In some cases we need to surrender back to the coastal areas and back to other river areas what it is that we should not be protecting—another way to let nature do its job.

"So some hard decisions have to come forward, and it's going to take a significant amount of time to mobilize the works, get them into place. But I think the key is first to recognize that we need to get ahead of flooding. It is, to my knowledge, the single most devastating and damaging thing to the people of the United States that we have to suffer within our confines. And with all of the technology that we have, we know how to start thinking about how to approach it.

"The problem is not the *way*. The problem is the *will* and the focus on prevention."

"New Orleans could be the test bed for the rest of the coastal portions of the U.S., and probably elsewhere in the world," says Van Heerden. "The Dutch have achieved it; sixty percent of their country's below sea level. Dealing with the soft soils of New Orleans, its unique geomorphology, and the fact that we've got these wetlands—that could be a real part of the protection. We could learn a lot as we try and sort this out that could be applicable elsewhere."

"I would agree with that," says Daniel, chair of the American Society of Civil Engineers, External Review Panel, and president of the University of Texas at Dallas. "And I think one of the lessons learned is that in critical life-safety issues like this, where infrequent but catastrophic events occur, like a major earthquake, we tend to forget as time passes. And history has also taught us that to make it work, you have to create very systematic, rigorous practices for continually assessing life safety. For example, in large dams—earthen and concrete dams—where failure would be catastrophic, we have federal programs in place to periodically inspect and reevaluate the safety of those structures. We have to be in this for the long run or we won't succeed."

CHAPTER TEN

THE MORAL IMPERATIVE

This is a moral issue, and it's a spiritual issue because how we respond to climate change is going to define what it means to be human.

— THE REVEREND SALLY
BINGHAM

Conservatives like to brand environmentalists as "tree huggers," a pejorative nickname for people who believe that the environment and the living creatures in it have a value that can't always be measured in dollars. Being labeled a tree hugger, to some, is a close second to being called a liberal, heaven forbid. But heaven, or those who believe in it, are now weighing in on the issue with a belief that is uniting environmentalists in a common cause.

At churches, synagogues, mosques, and temples around the country, religious leaders are bringing a message to their worshippers that is a decidedly green one: Stop global warming now. And it's not just a responsible thing to do, they say, but is more—something we're required to do in our role as stewards of the planet.

"The Buddhists believe that everything is interconnected, and if you harm some part of nature, you're harming yourself," says the Reverend Sally Bingham, environmental minister at Grace Cathedral in San Francisco. "The Muslims believe in a balance that God set up between nature and humans, and Christians believe that all things came into being through Christ; therefore, everything will be reconciled to God through Christ. And every major religion has a mandate for stewardship of creation."

As for the phrase in the Old Testament about people having "dominion over the Earth"? Bingham believes it has been misinterpreted. "*Dominion* does not mean 'dominate' or 'exploit.' Dominion is the same kind of perhaps dominion that God has over us, in that it's about care and love and stewardship. It's a mandate to be caretakers."

Bingham is founder of the Regeneration Project, which was started in 1993 and focuses on getting individuals to make more Earth-friendly energy choices. Congregations in almost two dozen states are active. "We ask congregations to join the program and then cut their carbon emissions—in other words, practice what they preach."

For example, in one Catholic church south of Detroit, a windmill, solar panels, and solar water-heating systems have helped the parish cut its energy use, its carbon dioxide (CO_2) emissions, and its energy bills. And in an effort to further spread the word, the organization has been holding free screenings of films on global warming, including Al Gore's *An Inconvenient Truth*.

"We like to have the priest or the rabbi walk down the aisle and be able to say this congregation is cutting their carbon emissions in response to climate change, but also it has an economic value too, because when you cut your energy use, you also are saving money."

The Reverend Bingham's religious fervor is more than just words. "The cathedral has compact fluorescent lightbulbs throughout. They have sensors in the bathroom so that if someone is in but leaves, it doesn't remain on. After five minutes, the lights go out. We have all

energy-efficient exit lights, the LED [light-emitting diode] exit lights."
And whenever a congregation gets with the program, the Regeneration
Project alerts it whenever their local utility offers rebates for energy-
efficient bulbs or is giving out free compact fluorescent lights. "We no-
tify all of our members that those perks are available."

Above all, the Reverend Bingham believes it is important not to
let politics get in the way. The environment is neither Republican
nor Democrat. It doesn't matter to Bingham what party Al Gore belongs to. And by the size of the crowds that turn out to see his film,
it doesn't look like they care, either.

"We advertised it not as a film about Al Gore but as a film about
the science of global warming. And to see the overwhelming response
was extraordinary, and that happened all over the country. That was
happening in Georgia, it happened in Arkansas. We showed this film
in every single state in the country, four thousand venues. We showed
it here at the cathedral, and we had expected fifty, perhaps at the most
seventy-five people, but three hundred people showed up to see this
film. And I asked the question: 'How many of you have seen it before?,'
thinking that maybe they were just coming in to see it for a second
time. But largely, they were people seeing it for the first time, and that
tells me that folks are hungry for the science."

And while Gore doesn't preach the gospel in his film, the con-
gregants see the message, clearly, in their own hearts and minds.
"They are beginning to understand it as a religious issue," says Bingham, who has been preaching about the environment this way for
more than ten years. "We have always preached and taught that this
is not a political issue. This is a moral issue, and it's a spiritual issue
because how we respond to climate change is going to define what it
means to be human today. How are we going to treat fellow walkers
of the planet? What kind of a future are we leaving for our children?
We can take this out of the political arena and make it a religious
issue."

EVANGELICALS IN THE FOLD

For the past 85 years, evangelical Christians have been making themselves heard, loudly protesting the teaching of evolution in biology classrooms, favoring the teaching of the biblical account of creation. Only recently have they become vocal in support of the environment, which is welcome news for the Reverend Bingham. "As far as the evangelicals, they are getting a huge amount of attention because they represent an enormous population in America. And the fact that they are now talking about "creation care" is quite wonderful. It may be just what we need to turn this whole thing around and get people all over the country that are sitting in the pews to respond to climate change in a religious and responsible way."

But Bingham says it won't be easy for evangelicals to unite with environmental scientists. "I know that just from the evangelical perspective, that's probably a huge challenge for them because evangelicals are taught from the very beginning to be suspicious of science—when they are talking about the difference between intelligent design and evolution. So now we're asking them to accept what the scientists are saying, and I believe that that's going to be one of the largest or biggest challenges for the evangelical community.

"I always say that the scientists are today's prophets. I mean, they are Hosea and Isaiah and Jeremiah, and we need to believe them. There is enough consensus in the scientific world that there's no reason for us not to accept what the scientists are telling us."

VOTING THEIR CONSCIENCE

Bingham doesn't believe in lobbying members of Congress. But she does support political initiatives and ballots in states that aim to clean up air pollution and foster energy conservation. She encourages people to vote their moral and religious convictions.

"We are called to serve each other and serve the poor. And the poor people are disproportionately affected by dirty-burning power plants because they most often live near them. And again, it's a social justice issue, and it's why religion needs to be involved in this discussion of solutions."

In fact, Bingham believes that her "interfaith power and light" program can serve as a model for other social issues, such as health, "because if we're going to be pro-life, we really believe that we ought to be pro-healthy-life. If there are 187 toxic chemicals in the cord blood of babies, we know that we are not producing healthy children. And I think that the religious community has a role here too. Look at abolition of slavery and women's right to vote and educate, and look at the civil rights movement. Once the religious voice was involved and people started seeing this from a moral perspective, things changed.

"When there's a major cultural change that needs to happen, if the religious voice is not there, it won't happen. And I believe that's going to happen now. I think our time is here."

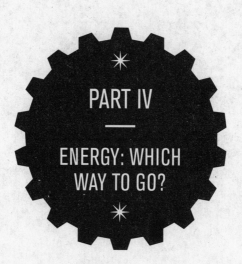

PART IV

—

ENERGY: WHICH
WAY TO GO?

CHAPTER ELEVEN

IT MAKES YOUR HAIR HURT

*Most problems have either many answers or no answer. Only a
few problems have a single answer.*

—EDMUND C. BERKELEY

*Georgia Institute of Technology's fuel cell aircraft flies above the track at
Atlanta Dragway. The pilotless vehicle flew for up to a minute at a time during test flights. Courtesy of Georgia Tech, Gary Meek Photo.*

Having a root canal done is less painful than talking about energy. In
the 35 years that I have been covering science and technology, no
topic has been more confusing to follow, more filled with politics, more
frustrating. Want an everyday example? Just look at the overnight

spikes in the price of gasoline. It jumps from one week to another. Oil companies make record profits. But if you ask why this is fair, why consumers should be paying so much more at the pump while the oil companies rake in the dough, they'll offer a jumble of confusing graphs and convoluted explanations that make sense only to other energy "experts."

In fact, I had a sense of déjà vu when in 2006 I asked a petroleum industry representative the same question I had asked in the 1970s, during the Arab oil embargo: I asked her to explain again, 30-plus years later, what accounted for that spike in gasoline prices. She offered virtually the same undecipherable explanation I had heard decades earlier. And when I told her it made no more sense now than it had then, she said, "That's my story." She might just as well have added, "And I'm sticking to it."

So as we peer into the energy crystal ball for a glimpse of where our energy future lies—will we be able to cut our "addiction to oil" with alternative energies?—your guess is as good as the experts'. Who knows? My advice is to listen to what Deep Throat told Woodward and Bernstein: "Follow the money." Energy is really *big* business. If the measure of success was the excellence of our products, we'd all be watching tapes on Betamax systems and using Macintosh computers. But money speaks louder than words and ideas.

Any change to our oil addiction will take years, if not decades, to bear fruit. And such changes, at least in the United States, may well hinge on what the person in the Oval Office feels are important to his or her goals. Remember, as a warrior battling the oil embargo, Jimmy Carter was very proud of the solar panels he put up on the roof of the White House. He started America down the road of energy conservation, with an effort to rally the troops against foreign oil with the battle cry "Conservation is the moral equivalent of war."

But no sooner had he gotten the ball rolling than along came the next president, Ronald Reagan, who, in one of his first acts, ripped the solar panels off the roof of the White House. Take that!

So much for alternative energy! So much for what *you* think we should be doing. Who knows what the next resident of the White House will think about George W. Bush's energy policy and what that person will do to change it. Incidentally, at just about the same time that Carter had realized that our reliance on foreign oil was an albatross, Brazil began a national program to convert its cars to burn alcohol instead of gasoline. Brazil has become so highly successful at doing that, planting and harvesting sugar cane (at the expense of the rain forest, unfortunately), that at the time this book was written, Brazil was virtually energy independent, no longer a fossil fuel–based economy. Thanks to flex-fuel cars. Brazilian drivers can choose to fill their tanks with 100 percent alcohol or a blend of alcohol and gasoline. And because they grow their fuel, they can replenish their source every year.

Imagine if the United States had gone down that road—and stayed on it—where its economy would be today.

Worth watching is the path that individual states take. Many of them, such as California and New York, are not waiting for Washington, D.C., to satisfy the dictates of the lobbyists. Instead, they are moving ahead independently, as California did to set automobile energy efficiency and greenhouse gas pollution standards. And as New York does to develop ethanol, not from corn—which yields very little net energy—but from willow trees. More on that later . . .

And Texas, of all places, has taken the lead in developing wind energy. By installing more and more wind turbines, Texas has now surpassed trendsetting California as the number one installer of wind-energy megawatts. If there is a story about individual states taking up the alternative energy slack, it is in wind energy. From New York to Kansas, from California to Massachusetts, state after state is realizing that it can slowly narrow its dependence on foreign oil by installing wind generators. And farmers are finding that they can make a huge profit by leasing their land to utilities that will install wind turbines on their acreage. Farmers can't get them built fast enough, and they can be installed very quickly, in a matter of weeks. The only problem the farmers face is what to do with their excess wind-produced electricity. With no national electric grid to tap into, farmers in the Midwest, in states such as Kansas, are left with surplus power potential. All dressed up with no place to go. Why not extend the grid to these homegrown energy producers? Ask your local politician.

So with your indulgence, I will attempt to tiptoe through the energy minefield and concentrate as much as possible on the science and technology. But be ready to ingest, along with the ABC's of energy, a hearty helping of politics and personal opinion. Energy is what drives everything, so it reaches into every aspect of our lives. Everyone has a stake, business and personal, in its future. But without a unified direction, without leadership to unite the disparate priorities, we will continue down a road of conflicting ideas and personal interests.

CHAPTER TWELVE

✳

FORESTS AND FIELDS OF ALCOHOL

The grass stretched as far as the eye could see, and hundreds
more miles beyond that. An ocean of grass—deep enough to
swallow a horse and rider—swaying and singing in the steady
wind of the Great Plains.

—OAK RIDGE NATIONAL
LABORATORY

In sports, they say, "Let's go to the videotape." In science, they say, "Let's look at the research." In both cases, it's a call to cut through the rhetoric and take a closer look at the facts. And the fact is that the United States is rushing headlong into turning its crops, especially corn, into fuel. Lester Brown, an economist who's president and founder of the Earth Policy Institute in Washington, calls it a stampede.

"The U.S. started the grain-to-ethanol program back in the late '70s after the two hikes in the world price of oil. And up until recently these biofuel programs—either ethanol or biodiesel—have been supported and driven by various incentives. For example, the 52-cent-a-gallon tax subsidy for ethanol production in the United States.

"But as the price of oil has climbed over the last year or so, it has become hugely profitable to convert agricultural commodities into fuel, and since everything we eat essentially can be converted either into ethanol or biodiesel, we're seeing an emerging competition now between the ethanol distilleries in this country, for example, and feed lots and food processors. And this is quickly becoming worldwide because of the prominent position the U.S. plays in the world food economy."

Market forces have taken over, says Brown, "so we're getting this stampede, almost a gold rush sort of mentality for investors wanting to build ethanol distilleries because they're so profitable. No one is really in control now. The market is driving the process."

Take Iowa, he says, our leading corn-producing state. "If all of the distilleries now in production plus those under construction and in the planning stages are completed, it will take the entire corn harvest of Iowa just to operate them.

"I don't think most people realize how much grain it takes to run an automobile," notes Brown, "but the grain required to fill a twenty-five-gallon SUV [sports utility vehicle] tank with ethanol will feed one person for a year. So if we're looking at filling the tank every two weeks or so, then we're looking at one SUV consuming the grain in the form of ethanol that would feed twenty-six people for a year. It doesn't take a lot of cars running on ethanol before you really begin to encroach on the food supply. I'm not saying let's close down the ethanol distilleries. What I am suggesting is that we need to take inventory. I think the first thing we need to do is for the administration to quickly do a tally of how many ethanol distilleries are in operation, under construction, and in the planning stage and then see how much grain that's going to take.

"The big risk is that we'll be using so much grain for cars in this country that there won't be enough for the rest of world, and the world depends heavily on us. So this competition between food and fuel is becoming very real, and the world is simply not prepared for it."

Secondly, says Brown, our best bet for developing ethanol is not to use food crops to produce it but to use other plants that we don't eat, such as "switchgrass or some agricultural residues or wood chips that will not compete directly with the food supply."

Food-based biofuels can meet but a small portion of our appetite for oil and gas for our cars and trucks. "If we were to convert our entire grain harvest in the United States into ethanol to run cars, it would supply something like sixteen percent of our total fuel needs." What's the solution? Make fuel out of plants—or parts of the plant— that we don't eat.

CELLULOSIC BIOMASS

The fibrous, woody, and generally inedible portions of plants are called cellulosic biomass. They contain cellulose, hemicellulose, and a component of the plant cell walls called lignin. It's all that woody stuff that we would normally throw away, and there's lots of it. It's the stalks and leaves of corn left behind after you take the cob. It's the tree limbs or vegetation removed to reduce forest fire hazards or the brush you clear from your ranch or backyard. It's the wood chips or sawdust from lumber- and paper-processing mills that normally get thrown away. It's wood or paper products and the grass clippings and food scraps you put into your compost heap.

All of it can be converted into a variety of high-value fuels: ethanol, biodiesel, methanol, hydrogen, or methane. And much of it can be renewed—grown again—and harvested in fields or even vacant lots.

"If you get your ethanol from cellulose, it's just a really exciting big winner on both sides," says Dan Kammen, director of the Renewable and Appropriate Energy Laboratory and codirector of the Berkeley Institute of the Environment at the University of California, Berkeley. "We have a huge amount of biowaste in this country." For example, sawmills, says Kammen, waste a lot of wood. About a third of the material from a tree ends up on the cutting-room floor.

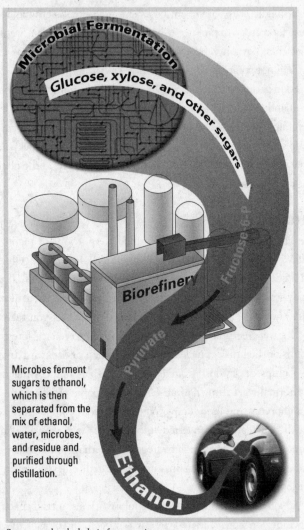

Steps to make alcohol via fermentation.

"All that stuff is potential crop matter. Lots of the residues from different farm products are also potential products. And so we've got already in this country a very large resource of unused cellulose that we could make into ethanol without even converting one more acre of land into a bio-energy crop. We haven't even begun to explore how much of our current waste stream we could put into this."

Turning waste cellulose into clean-burning energy offers an added bonus of cleaning up the air. Kammen says lots of that waste cellulose in the form of husks and shells that might otherwise be burned can be turned into ethanol, "and so in California's central valley, where there are real air-pollution problems from burning rice shells and the shells from walnuts," there is the potential to eliminate all that air pollution.

And if you wanted to grow crops specifically to turn them into ethanol, other plants yield a whole lot more energy than corn does.

FORESTS OF ALCOHOL

Wood pulp (shown here) and other waste plant materials are an abundant source of feed stock for producing cellulosic ethanol.

Even with all the waste cellulose available to us, we could increase the amount of biofuels by growing it. "Like switchgrass, which can grow in prairies, says Kammen, "as well as some of the fast-growing trees."

Trees? Absolutely. Try poplar and willow trees. The state of New York, for example, is experimenting with turning northern hardwoods, willows, into alcohol. "In New York State, we've got 18.5 million forested acres. And much of that acreage is made up by our northern hardwoods," says Dr. Carol Murphy, executive director of the Alliance for Clean Energy New York. "There's a series of compounds in those hardwoods called hemicellulose. We're looking at how we can extract the hemicellulose from the northern hardwood chips, convert the hemicellulose into five-carbon sugars, and ferment the five-carbon sugars into ethanol."

And what makes those trees so enticing? Remember our energy numbers? Remember that corn ethanol gives you a return of 1.25

energy units for each energy unit you put into growing, harvesting, and turning the corn into ethanol? Well, how about *these* numbers? "In the case of our northern hardwoods, for every unit of energy in, you get fourteen units, to maybe sixteen or eighteen units" out. That's more than ten times what you get from corn. What a difference!

New York State has 600,000 acres of land not currently engaged in agriculture. There are another 600,000 acres of land that are probably a little too wet to be used for crops but that all would be just great for growing willows. "We can literally grow our biomass on that 1.2 million acres of land, give a return to our agricultural community, and help wean us off the petroleum base, particularly for transportation fuels."

Just as forests are grown and harvested for plywood pulp, forests could be grown and harvested for fuel. And willows are an ideal crop, says Murphy. "We can start with a six-inch sprig and literally in three years it's thirty feet tall." So while a plywood forest may take 10 to 15 years to mature and harvest, a willow forest takes just three years. And Murphy thinks that New York can get this project underway in as little as two or three years.

But while growing energy forests may be part of an overall energy future, it won't serve as the sole solution. "Some estimates have been made that to offset a million barrels of oil, you need basically a billion tons of biomass. There's no question that we need to look very carefully at some of those things. But equally there needs to be balance in our approach to renewable energy. There is not going to be any single solution. It's going to require a number of different sources of renewable energy using biomass as a source, indigenous sources of biomass. We need to take a balanced approach. The *total* solution to our problem certainly isn't willow and certainly isn't ethanol from willow."

SWITCHGRASS: BACK TO THE FUTURE

Switchgrass once covered the American plains as far as the eye could see. Grown easily anywhere, it is a perfect plant for producing ethanol. Courtesy of the U.S. Department of Energy.

A key answer to bio-fuels might be turning to a plant once so abundant in the Great Plains that it stretched as far as the eye could see, and even then, for hundreds of miles more. It's called switchgrass. And it is one of the few plants singled out for mention by President George W. Bush when he talks about renewable biofuels. This perennial's potential for becoming a major source of alternative energy has even made the federal government's Oak Ridge National Laboratory (ORNL) wax poetic about the endless acres of switchgrass that used to grow wild on the American frontier:

> The grass stretched as far as the eye could see, and hundreds more miles beyond that. An ocean of grass—deep enough to swallow a horse and rider—swaying and singing in the steady wind of the Great Plains. The American prairie—tens of millions of acres—once looked like this. But that was centuries ago, before the coming of the white man, the railroad, and the steel plow. Today, corn and beans hold sway, and the remnants of America's tallgrass prairie are confined mostly to parks and preserves. Now, though, in research plots and laboratories in the Plains states and even in the Deep South the seeds of change are germinating. The tall, native grasses of the prairie, so vital to our land's ecological past, may prove equally vital to its economic future. Such grasses once fed millions of bison. Soon, grown as energy crops, they may help fuel millions of cars and trucks, spin power turbines, and supply chemicals to American industries.

Plants like switchgrass don't need to grow on high-priced farm-land but can thrive on marginal lands. They don't need expensive, energy-intensive fertilizer or pesticides. They thrive in dry soil and their roots can reach deep down for a drink. Like willows, switchgrass grows quickly; it can stand 10 feet tall after one season. It processes sunlight efficiently, and turns photons and CO_2 into rugged cellulose.

And because it's been around for millions of years, switchgrass has learned how to live in just about any soil and in harsh climates. It can be grown in fields and harvested and baled like cotton, needing to be replanted only once every 10 years or so. Many plants deplete the soil as they grow, but not switchgrass. It adds organic material to the soil as it grows, and its extensive root system fights winter erosion.

And "buffer strips of switchgrass, planted along stream banks and around wetlands," says ORNL, "could remove soil particles, pesticides, and fertilizer residues from surface water before it reaches groundwater or streams—and could also provide energy."

Test plots of switchgrass at Auburn University, says ORNL, have produced up to 15 tons per acre, equivalent to 1,150 gallons of ethanol per acre per year. And researchers are genetically engineering the plant to become even better, making it even more adaptable to a wider range of growing conditions. So not only is the yield better per acre but also the plant will be able to grow under more adverse conditions. I could go on singing the praises of switchgrass, but you can read about it yourself at http://bioenergy.ornl.gov/papers/misc/switgrs.html.

Why not have farmers change crops, from, let's say, tobacco—the crop of yesterday—to switchgrass, the crop of tomorrow? "I'll grow anything," a farmer once told me, "as long as I can make a profit."

"The reason corn has become so popular is because it's there and we're producing it in huge quantities," says Brown, "and there's strong support from farmers to convert part of that crop into ethanol. But the willow trees and the fast-growing hybrid poplars are two of the strongest candidates for cellulosic ethanol production, along with switchgrass."

Here's my vision of our energy future: fields of switchgrass, forests of willows and poplars, and planted among them, rows and rows of wind turbines, all creating a future of clean energy independence. But that vision may have to wait a bit because the technology for turning cellulose into fuel is not quite ready for prime time, says Brown. "We're probably at least five years away from technologies to convert either willow trees or switchgrass into ethanol on an economically competitive basis."

In addition, other alternatives to ethanol may be more productive, says Brown, such as a move toward plug-in electric cars powered by electricity produced by wind turbines. "If you take a car like a Toyota Prius, which is the most widely sold gas–electric hybrid, and if you add a second storage battery and a plug-in capacity, then we can do most of our short-distance driving—commuting, grocery shopping, and so forth—almost entirely with electricity. And the idea that we now have the technologies and an abundance of wind resources that would permit us to run our cars largely on wind energy is, I think, very exciting. Especially when you realize that the costs of the wind-electricity equivalent of a gallon of gasoline is less than a dollar a gallon. There will be no source of ethanol, even cellulosic ethanol, that'll be able to compete with the equivalent of a dollar-a-gallon wind energy." Wind turbines can produce electricity at under five cents per kilowatt-hour, making your plug-in auto cheaper to run than a gas-powered model. And as turbine technology progresses and the number of installed wind turbines grow, the costs can come down even more.

ALCOHOL VERSUS DIESEL: A COMPARISON

Given the headlong rush to build ethanol plants, some of the world's top economists and agriculture experts are wondering if it makes any sense to grow crops just to turn them into alcohol. For example, one study shows that when you take into account the economic, environmental, and energy costs and benefits, biodiesel is a better choice than corn ethanol as a fuel.

Research teams at the University of Minnesota and St. Olaf College studied alternative biofuels—that is, fuels made from plants as opposed to oil, coal, and natural gas. To be a viable alternative energy source, they wrote in the *Proceedings of the National Academy of Sciences*, a biofuel should produce more energy than it takes to grow and process, it should have "environmental benefits," be able to compete economically with other fuels, and "be producible in large quantities without reducing food supplies." Using those criteria, they compared the life cycle of corn used to make ethanol with the life cycle of soybeans used to make biodiesel.

Their results were startling. Ethanol returns 25 percent more energy than it takes to put into it. So if you put 100 units of energy into growing, harvesting, and turning corn into alcohol, you get a yield of 125 units of energy. But biodiesel yields 93 percent more. Put 100 units of energy into growing, harvesting, and turning soybeans into diesel and you get 193 percent of energy out of it. That's a tremendous energy-saving advantage over corn.

What about the environmental impact of each one? If biofuels are used instead of fossil fuels, "greenhouse gas emissions are reduced twelve percent by the production and combustion of ethanol and forty-one percent by biodiesel," according to the research teams. Wow. "And pollution-wise, biodiesel also releases less air pollutants per net energy gain than ethanol."

Why the big difference? The advantage of biodiesel is that it takes less energy to grow soybeans than it does to grow corn. And it's a lot more efficient to convert soybean feedstocks to create biodiesel than it is to convert corn to ethanol.

On the other hand (and there *always* is another hand), even if you were to dedicate all the corn and soybean production in the United States to making biofuels, they "would meet only eleven percent of gasoline demand and eight-point-seven percent of diesel demand," the researchers wrote. Clearly energy conservation and alternative energies have to be part of the solution.

Switchgrass could be one of those energy alternatives. It compares much more favorably with soy. But soy is already an established crop, the second largest in the United States, right after wheat. On the other (third?) hand, growing soy for fuel raises the same problem that comes with growing corn for fuel: You drive up the price of food. Which might be politically unacceptable. By growing switchgrass, you do not face a choice between food or fuel.

CHAPTER THIRTEEN

THE NUCLEAR OPTION

*It is not too much to expect that our children will enjoy in their
homes electrical energy too cheap to meter . . .*

—LEWIS STRAUSS, CHAIRMAN
OF THE U.S. ATOMIC ENERGY
COMMISSION, 1954, SPEAKING
ABOUT THE FUTURE OF NU-
CLEAR ENERGY

It wasn't too long ago, after World War II, that the U.S. government
had such high hopes for nuclear energy that it thought nuclear power
would be so plentiful and cheap that it would be given away. Of
course, that never happened. Nuclear power did get a foothold in
this country, but the meltdown of nuclear fuel at the Three Mile Island power plant in Middletown, Pennsylvania, in 1979 put the lid
on nuclear power development in this country.

But not in Europe. France gets about 75 percent of its electricity
from nuclear power. Here in the United States, nuclear power usage
is less than a third of that. France has brought 58 nuclear plants online since the 1970s. The United States hasn't ordered any new
plants since the Three Mile Island accident.

But that may all be changing. President George W. Bush has said

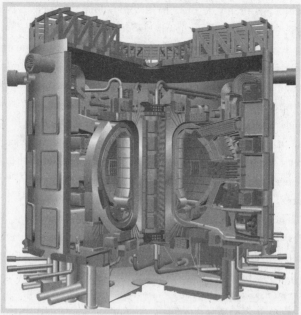

Nuclear fusion, perhaps the ultimate source of clean and abundant nuclear energy but so far, impossible to produce economically: This is a cross section of a design of a fusion reactor called the Tokamak. Courtesy of the U.S. Department of Energy.

that he wants to see an increased emphasis on nuclear power in the United States, including starting construction on new plants by 2010. But is this country ready for nuclear power? Has the technology improved to the point where concerns over issues such as safety and waste disposal can be addressed? In the land where the phrase "Not in my backyard" strikes fear in the heart of politicians, is there a political will to build new plants?

But ironically, we are seeing some environmentalists who once were rigidly opposed to nuclear power now saying that compared with global warming, nuclear energy poses a much smaller threat. Take the ultimate tree hugger, Dr. Patrick Moore, cofounder and former leader of Greenpeace. He has helped create CASEnergy

Coalition, a group devoted to helping promote nuclear power. "There is a great deal of scientific evidence showing nuclear power to be an environmentally sound and safe choice. A doubling of nuclear energy production would make it possible to significantly reduce greenhouse gas emissions while increasing our energy supply."

Other environmentalists are not convinced. But interestingly enough, where they used to argue that nuclear power was risky because of the threat of a meltdown, as happened at the reactor at Three Mile Island or in the explosion and fire at Chernobyl, they no longer present those arguments. Rather, they argue three other points. The first is economics.

"The rub for nuclear power today, particularly in the United States, is that it's uneconomical compared to alternative energy sources for producing electricity." That's Dr. Tom Cochran's argument against nuclear power. He's director of a nuclear program at the National Resources Defense Council.

"Wall Street has no confidence in nuclear power, and although a lot of people are talking about it, nobody's really ordered a reactor yet. And I hope the government doesn't get into the business." That's Dr. Arjun Makhijani, president of the Institute for Energy and Environmental Research in Takoma Park, Maryland.

The second problem is waste storage. Where will we put all the spent nuclear fuel rods that nuclear plants leave behind? "The real issue is what do you do with this material over the long term, over the tens of thousands and hundreds of thousands of years that it will remain radioactive?" That's Dr. Kevin Crowley, a geologist and director of the Nuclear and Radiation Studies Board at the National Academy of Sciences.

The third is terrorism and nuclear proliferation. In a nuclear society that relies on large quantities of nuclear materials, there is always the threat that terrorists might get their hands on radioactive materials and make a crude nuclear bomb or even a nonnuclear but deadly "dirty" bomb.

"They're into 'Will it explode?' And they don't care if it's one kiloton or ten kilotons," says Makhijani.

There is also the possibility that "peaceful" countries, when faced with newly nuclear hostile nations, may find a way of turning their nuclear power reactors into breeding grounds for nuclear weapons.

THE FRENCH CONNECTION

Searching for answers to these problems, technologists look for similar scenarios where nuclear power appears to be working well: France. France is the poster child for nuclear energy. The vast majority of the electricity in that country comes from nuclear power. Why can't we just do what the French do and build more nuclear reactors?

"In order to address it to the level of France, we need about seven hundred or eight hundred nuclear power plants here in the next fifty years, if you want seventy-five to eighty-five percent in this country," says Makhijani. "It's much bigger than France with a much bigger electricity sector. That's about maybe two a month, or three every two months, for the next forty years. Not an achievable level." And even if you could build that many plants and run them for the rest of the century, says Cochran, where would you put all those spent, highly radioactive nuclear fuel rods? "You would need something like another fourteen or so Yucca Mountain–size repositories," says a skeptical Cochran, referring to a site in Nevada that is being studied as a place to store the spent nuclear fuel rods.

THE STORAGE PROBLEM

Even if you build only one new nuclear reactor, you're faced with a problem that has not been solved since the first reactor was patented in the 1950s: where to store the highly radioactive, highly lethal, nuclear waste—the used-up fuel rods that are taken out of the reactor?

"The nuclear waste disposal issue has been characterized by some as the elephant in the living room," says Crowley. "We've had a waste disposal problem since the late 1950s when the first commercial nuclear reactors began operating. We do have a short-term solution to the problem, and that is basically to store it at the sites at which it's generated. So since the late 1950s, the spent fuel that has been produced by operating nuclear reactors has been stored in the large water-filled pools called spent fuel pools at the nuclear reactor sites.

"And as those pools have begun to reach capacity, at some sites, nuclear power plant operators have taken some of the older fuel out of the pools and put that into large, heavily shielded structures called dry casks. And dry cask storage facilities are now beginning to be built at many plants."

Crowley says there's a general scientific consensus that the storage at plant sites can be carried out safely for decades if appropriate attention is paid to managing the waste. "The real issue is what do you do with this material over the long term, over the tens of thousands and hundreds of thousands of years that it will remain radioactive?"

A solution was proposed back in 1957, by the National Academy of Sciences: bury the wastes deep underground. The place currently under consideration as a deep burial site is Yucca Mountain, a ridge line in Nevada. It's been selected from a handful of potential sites to be the final resting place for high-level nuclear waste, the by-products of nuclear power and nuclear bombs, that are stored now in those casks and pools at 126 sites around the country. The U.S. Department of Energy has been studying the site since 1978. Various political, legal, and scientific controversies have flared up over the decades. Some claim that the site is not suitable—read: not safe—for the storage of nuclear wastes for tens of thousands of years. Other efforts come from the state of Nevada, trying to get the storage site moved elsewhere. The Department of Energy, in 2006, asked Oak Ridge Associated Universities to analyze the scientific basis for storing the wastes there, and to arrive at a judgment about the safety of Yucca Mountain as a storage site.

"The current plan is to submit a license application to the Nuclear

Regulatory Commission by the end of 2008," says Crowley, "and then to begin operation of the repository no later than 2020. At this point, we're still waiting for the Environmental Protection Agency to issue health and safety standards for Yucca Mountain. And until they do that, the Department of Energy cannot complete its license application."

Once again, scientists are wading into unfamiliar territory, trying to make decisions about events they can't possibly foresee thousands of years into the future. Crowley continues, "This is a first-of-a-kind endeavor, and it's a very technically difficult endeavor, because the Department of Energy has to demonstrate with a high degree of confidence that the repository that they would build and eventually close at Yucca Mountain could contain the radioactive material for very long periods of time. And it's really establishing that long-term confidence in the performance of the repository that is the challenging technical issue."

Has is been established? "It has not been established yet," says Crowley. But . . . "I think that there's a strong consensus in the scientific community that it's possible to establish that basis."

The debate over the future of Yucca Mountain and the disposal of nuclear waste can get as heated as the nuclear fuel itself. Even some people who believe in the technology believe that the waste can and should be buried, deeply underground, do not believe that Yucca Mountain is that place.

"Yucca Mountain just happens to be the worst single site that has been selected or studied in the United States," says Dr. Makhijani, who says he has studied the repository problem for the past 25 years. "I say this as a supporter of a repository as a least-worst solution to a very big problem that we've created." He is most vocal about the potential problem for drinking water being contaminated thousands of years from now.

"In 1983, the Department of Energy-commissioned study from the National Academy of Sciences projected drinking-water doses. The department's own studies, its own contractors, have published graphs and charts" showing that according to the standards that the National Academy of Sciences has advocated for more than twenty years, "Yucca Mountain would not meet existing repository standards. In fact, rules have been changed four different times to accommodate Yucca Mountain because Yucca Mountain simply can't meet the standards. And I say this—unlike many environmentalists who don't support repositories, I do think we need one. I think we've rushed into site selection."

Rushed? It's been studied for decades!

"Unfortunately—and here I sympathize with the utilities and even with the Nuclear Energy Institute—the government has wasted most of this money. It's wasted it on a site that it knew could not really meet the standards. Instead of going to a new repository, we created another standard. I call it the double standard."

Even if Yucca Mountain were certified as a repository, certainly somewhere down the road another Yucca Mountain would need to be opened; more sites would need to be considered. How many more would we need? Crowley says the jury is still out on this one. "If we proceed down the path that we're going down now, which is to simply put the spent fuel into a repository, Yucca Mountain, the currently legislated limit is about seventy thousand metric tons. At present, we have about fifty-five thousand metric tons of commercial spent fuel in the United States. So we will soon fill up Yucca Mountain at the present rate of generation of commercial spent fuel, about two thousand metric tons a year. However, the capacity at Yucca Mountain is a legislated limit, and Yucca Mountain can be expanded if Congress would choose to do that. Some studies suggest that it could be expanded by a factor of five to ten in capacity."

Another possibility is shrinking the size of the nuclear waste. Scientists are looking into the future to do that by technologies that remove the "unburned" nuclear material that is normally thrown out with the spent fuel rods.

"It's like having a log in your fireplace, and we burn three percent off of one side, three percent off the other side. Then we take that log and throw it into a mountain and bury it," says Judy Biggert, Republican congresswoman from Illinois and chairperson of the House Subcommittee on Energy. "With recycling and reprocessing, we'll be able to reduce not only the size of it but also the toxicity of it so that it won't last so long in its radioactivity state." The idea is that if you can shrink the size of the waste, you won't have to find more disposal sites, says Biggert. "It's been said now that if we have the recycling, we won't need another Yucca Mountain until the next century, which is quite a far way away." This is the school of thought that says that if we put off the problem long enough, someone will invent a technology in the future that will make it go away. It's quite common among politicians who like to leave very difficult problems for our children to solve. See Social Security. See global warming. You can supply your own problem here. . . . But I digress.

In this case, scientists have already found a way of shrinking the waste by recycling and extracting the useful material. Which brings us full circle, back to France. "Everybody points to France, because it's the center of the reprocessing industry in the world," says Makhijani. "Well, I've studied the French nuclear industry quite a bit. They don't actually use all of the plutonium they separate. There's eighty tons of separated plutonium stored at La Hague. There's eighty tons of separated plutonium stored at Sellafield in Great Britain. This is an invitation to proliferation problems. La Hague supplies separated plutonium to Japan, which hasn't used a single ton of it as yet, and a couple of years back, their Labor Party leader, Mr. Ozawa, suggested that Japan should use or could use, if China starts acting up, their commercial plutonium for thousands of bombs."

Biggert agrees. The Japanese produce pure plutonium in their recycling program, which makes it useful for making bombs. But the American scientists at the Argonne National Laboratory, which happens to be in her district, have developed a recycling process that does not make pure plutonium.

"It's mixed with other nasty elements, americium, neptunium, and curium. And these are things that would make it practically impossible to sort out any plutonium for proliferation, so that there would be no chance of nuclear proliferation."

Makhijani does not agree. He is familiar, he says, with the Argonne separation method called electrometallurgical processing. He says she's right about the end product but wrong about its impact.

"It's true that it has neptunium and curium and americium mixed with it, so that no nuclear weapons state would want to use that plutonium. However, americium-241 and neptunium-237 are also fissile materials. And terrorists, for instance, or states who don't have nuclear weapons–useable materials could quite easily make nuclear weapons, especially if you're not worried about radiation doses."

PLANT SAFETY

Plant safety—whether nuclear material will escape from a plant during an accident such as the one at Three Mile Island—used to be at the top of the list of nuclear energy worries. It no longer is, but it's still there. To hear lifelong nuclear watchdogs like Tom Cochran talk, the change is almost palpable.

"In the United States, today, reactors are safer than they were a couple of decades ago. We haven't had a core melt accident since Three Mile Island." That is not to say there is no cause for worry. "In 2002, we had a major precursor at the Davis-Bessie plant, where there was discovered a football-sized hole had corroded in the reactor head because of lack of regulatory oversight over that reactor and utility oversight." Which leads Cochran to believe that the biggest challenge to plant safety comes from whether a good "safety culture" is instilled at the site. He says that the safety culture at U.S. plants has improved.

Makhijani is not so sanguine about the prospects. Take the Davis-Bessie plant corrosion. "We were not far from a very, very serious accident. And it turns out the French had warned the Americans about this problem. There were corrosion problems at a number of plants that were similar, from boric acid. And the Nuclear Regula-

tory Commission [NRC, which regulates the nuclear industry] simply wasn't paying attention. What we've got is a situation where the
NRC is much more lax than in the '80s. I think [that since the early
'90s] we've been asking for trouble. The industry isn't necessarily
safer. The Nuclear Regulatory Commission has a smaller technical
staff, and it is allowing self-inspection and self-regulation, and there
have been many problems where we've been very, very close."

NUCLEAR ENERGY IN THE AGE OF TERRORISM

Terrorism—individual and state-sponsored—has added an extra dimension to the nuclear energy equation. (Consider that fuel recycling problem we mentioned a short while ago.) When the Soviet
Union fell, there were fears that some of the nuclear material in the
warheads of the thousands of Soviet missiles might find its way into
the hands of terrorists, either by being stolen or by being sold. That
fear has not gone away; it has only been compounded by the fear
that a worldwide move toward nuclear energy might make it easier
for terrorists to obtain the uranium and plutonium fuel that powers
these reactors. With such radioactively lethal material—and it
doesn't take a whole lot of it—terrorists might fashion crude nuclear
weapons or dirty bombs.

In an effort to keep nuclear material out of the hands of countries
that might want to create nuclear weapons, President George W. Bush
in 2006 proposed a plan to "lease" nuclear fuels to countries that have
not signed the nonproliferation agreements, such as India. In this sce-
nario, the nuclear fuel would be watched from cradle to grave, as it is
sent to countries to fuel their energy-producing nuclear reactors and
retrieved later as spent fuel. The spent fuel is reprocessed so that the
plutonium in it can be used again in a nuclear reactor but not used for
nuclear weapons. Bush called it the Global Nuclear Energy Partnership,
and it is officially described as enhancing "energy security, while promoting nonproliferation. It would achieve its goal by having nations
with secure, advanced nuclear capabilities provide fuel services—fresh

fuel and recovery of used fuel—to other nations who agree to employ nuclear energy for power generation purposes only."

The partnership plan calls for hundreds of millions of dollars to be spent on this effort. New kinds of fuel-retrieving reactors must be developed too. But not everyone is impressed. "The Bush global nuclear energy program really puts nuclear energy on a much more slippery slope, much more proliferation prone," warns Makhijani. "It creates a kind of a nuclear apartheid in the world. It talks about fuel-cycle countries, in a very polite way, which can be trusted with plutonium, and then other countries which are reactor countries. This is like a unilateral amendment of the nonproliferation treaty, which guarantees its parties that they have the 'inalienable right' to nuclear power. If you promote the separation of plutonium, and promote nuclear energy, you're going to wind up with uranium-enrichment and reprocessing technologies throughout the world. Not, I think, a very pleasant future to look at."

Furthermore, Makhijani says that it's erroneous to assume that the reprocessing of spent nuclear wastes will not result in weapons-grade material finding its way into the hands of terrorists.

"This global nuclear energy program of reprocessing plutonium separation, being promoted as proliferation resistant—when you can make bombs out of the stuff—is really not a sensible program, and many advocates of nuclear energy are quite skeptical.

"Terrorists won't care if they're [using] impure plutonium, so long as you can make a chain reaction out of it. And they don't care if it's one kiloton or ten kilotons. A nuclear fizzle is several hundred tons of TNT equivalent. And I think for a terrorist fizzle, it would have a huge effect."

He points out that there is no guarantee that the countries will not take their apportioned plutonium and turn it into nuclear weapons. "Look at France, the sort of model country for plutonium separation. They reprocess for foreigners." And one of those clients is Japan, which uses separated plutonium in its commercial sector. In

2002, says Makhijani, Japan's Labor Party leader said that if China, its historic rival, gets too uppity, "Japan should make thousands of nuclear weapons from its commercial-sector plutonium. Now, that's the real world we are in, once we start separating plutonium. Commercial plutonium was said by Japan to be not suitable for nuclear weapons only ten years ago, but suddenly they're saying they could make thousands of weapons out of it."

Tom Cochran is even more critical. "India's first nuclear weapon was produced using plutonium from their research reactor that was supplied under the 'atoms for peace' program. It was reprocessed in a reprocessing plant that was supposed to be part of their breeder reactor program. So the United States, under the Ford and Carter administrations, stopped commercial reprocessing in this country.

"And when President Reagan tried to renew it, there was no utility interest because it was uneconomical, as it is today. So there's no economic reason to move ahead with reprocessing at this time. It's dangerous if it's pursued. And nonweapons states that are of concern, such as Iran? I mean, the last thing you would want to have in countries like Iran and North Korea are reprocessing plants and large stocks of spent nuclear power reactor fuel, because that's a quick access to nuclear weapons capability."

THE PEBBLE BED TO THE RESCUE?

What if instead of a giant nuclear reactor, you could build a smaller, much more efficient, modular unit that could be expanded at will and could compete with very large nuclear power stations that cost a huge amount of money? And on top of that would be much safer to operate and almost impossible to melt down? That's the promise of a technology, decades in the making and still under development, called the pebble-bed reactor.

"Imagine a bubble gum machine full of big round bubble gum balls," says Dr. Andrew Kadak, professor of nuclear engineering at Massachusetts Institute of Technology (MIT). Only instead of being round, these gum balls are cylindrical, little "pebbles." Each one of these little pebbles is packed with a load of radioactive fuel, "a cue ball containing, inside, ten thousand tiny little microspheres of uranium." They drop into the top of a hopper and by "gravity are circulated and discharged from the bottom and then pneumatically reinserted in the top and could be operated without refueling or shutting down for about five years. Draw this little mental diagram and you'd have the pebble-bed reactor," says Kadak, completing the picture. Kadak and his colleagues at MIT are working on developing this reactor, first demonstrated by the Germans, with other countries in the hunt. "The Chinese and the South Africans are in fact now in the process of licensing for construction two of these demonstration plants."

You'd have to say that a nuclear power plant, built like a candy dispenser, is highly unique, to wax redundant. When compared with your standard hot-water reactor that steams water vapor out of its cooling towers down the block, the pebble bed is a radical design. Consider this: In addition to the nuclear jaw-breaker fuel, the reactor has *no water*. Nope. Instead, the coolant is helium, high-temperature helium gas that's circulated instead of water. "This helium gas is an inert gas, which means it doesn't get activated or corrode materials, which is an attractive feature," compared with the water reactors that are eating nasty holes in the reactor walls, Kadak says. The pebble-bed reactor also promises to be much more efficient, achieving efficiencies as high as 50 percent.

And that gas design allows for flexibility. "We can either take the helium and put that helium to a gas turbine, which makes electricity directly, or we can process it through a heat exchange and either make steam as the Chinese are doing or put this device, this intermediate heat exchanger, on the nuclear side of the plant, [so] you're able to apply that heat to many, many different applications," Kadak says. There is even speculation that the reactor can be made to produce hydrogen, creating the basis of a hydrogen economy driven in part by nuclear reactors of the future.

Oh. And speaking of walls. This reactor does not have them, at least not the superthick concrete containment walls that prevented the contaminated water from leaking out of Three Mile Island's melting nuclear reactor. No walls, because there is no water. The proponents of this design say the reactor is "meltdown-proof" because the reactor's design does not let the fuel get hot enough to melt. The reactor is designed to shut itself down without human intervention if the coolant fails.

Wow. Does this idea sound hot! So to speak. Of course there are the minuses. "You're doing away with essentially what has served as the foundation of the safety margin in the reactors that we have," says Makhijani. "Most of the severe accidents that have happened so far have happened in graphite-moderated reactors. Reactors that can catch fire, because they've got carbon, Chernobyl, Winscale, and so on. And pebble bed is a graphite-moderated reactor." It's the graphite that slows down or moderates the speed of the nuclear particles in the reaction, so that the reaction can proceed.

Kadak agrees that the potential fire threat is something to be studied. "There's a lot of discussion about what actually happened at Chernobyl, whether it was the graphite burning or the zirconium cladding burning. But clearly, the graphite supported the combustion, whatever fuel that was there, that was burning.

"We're now doing studies at MIT and the Germans have done numerous tests on these things called air-ingress accidents, to assess whether there is a graphite-burning issue." The studies to date have indicated, he says, that there is no graphite-burning issue for this particular design and for this particular configuration. "But the Nuclear Regulatory Commission will be the ultimate arbiter on that point."

Now for the real naysayers. "This is a nice research project," says Cochran, "but it isn't going to solve the global warming problem, because pebble-bed reactors don't show any signs of being cheaper than the conventional reactors we have here in this country. And in fact, the only utility that was supporting R&D on this reactor in the United States backed out when the CEO resigned from the company."

In fact, even the White House has been reluctant to spend money on developing these reactors, spending money instead on more conventional nuclear reactors. That's no mistake, says Kadak. "From a practical standpoint, the next fleet of reactors will have to be these advanced light-water reactors, which have significant improvements in overall safety system performance. And the pebble-bed reactor, the first two real commercial plants are going to only be demonstrated if they come online as scheduled by 2011, 2012. So once these reactors come online, then the rest of the industry and the United States for sure can look at this and say whether or not this is something that they would like to invest in for the longer term." Those two reactors are the ones South Africa and China will be pioneering.

THE ECONOMICS: FOLLOW THE MONEY

In the final analysis, it's all about money. One can't speak about nuclear power in a vacuum; it is always compared, economically, to other alternative energies. "The economics is going to be the driver, relative to what new technology is developed, whether it be wind, solar, nuclear, or coal, says Kadak, reflecting the consensus of just about everyone in the energy business. "Does nuclear power offer an economic advantage over the other alternative energy options?"

That's where the debate begins. Historically, nuclear power has not been very economical. "Quite frankly, until the beginning of the '90s, we did not operate the plants as well as we do today," says David Modeen, vice president of the nuclear division for the Electric Power Research Institute. "We've had a steady increase in the efficiency and the effectiveness today where the fleetwide average is ninety percent efficiency. It's the lowest-cost-based generation by all metrics both industry and government."

"They keep saying it's very economical," rebuts Makhijani, speaking of the nuclear power industry. "But Wall Street doesn't agree with that. If you include the capital cost, which is most of the cost, then government subsidies still appear to be necessary half a century after the start of nuclear power. Otherwise they'd go and order them." And in fact at the close of 2005, the White House authorized the federal government to issue more than $10 billion for the construction of a half dozen nuclear power plants.

"But how far is that going to take us?" Not much of an impact on global warming, says Cochran, or on energy independence. "Our growth in electricity requires twenty, thirty plants every year, and we have six plants in ten years; it's really a few drops in the bucket. There are better alternatives that will impact the global-warming pollution problem faster and cheaper and safer than subsidizing new nuclear plants."

Putting the final coda on the French nuclear model, Makhijani sums up the feelings of many energy experts in the United States. "Nobody's lining up to really build a lot of nuclear power plants. And to say that nuclear power plants are going to be the answer to the electricity sector's woes and to global warming is—so far, at least—in fantasy land."

Referring to Cochran's emphasis on alternative energies, Kadak says, "He's quite right that the economics is going to be the driver, relative to what new technology is developed, whether it be wind, solar, nuclear, or coal. When you start looking at the economics of new nuclear plants, it's all quite speculative in the sense that we haven't built one in the United States for such a long time. We really don't know what our current costs are, and those people who are promoting the construction of these plants believe that they can build them much cheaper" than what studies like those done at MIT show. "So the proof will be in the pudding."

"What is in short supply is not energy sources," says Makhijani. "Wind can offset CO_2. Solar can offset CO_2. Nuclear can offset CO_2. What is it going to cost? Nuclear is among the more costly sources."

IS COAL STILL KING?

Not all the coal that is dug warms the world.
—MARY H. JONES

In 2003, President George W. Bush announced that the U.S. government would spend $1 billion to finance a 10-year demonstration project "to create the world's first coal-based, zero emissions electricity and hydrogen power plant" called FutureGen.

"We are the OPEC of coal. We have more energy in the form of coal here in the United States than the Middle East has in its entirety in oil," says Joe Lucas. As executive director of Americans for Balanced Energy Choices, Lucas can be called America's coal ambassador. Its coal guru. Perhaps its biggest advocate. His organization may say "energy choices," but his choice for energy is certainly coal. And if you look at just the numbers, that makes a lot of sense.

"Today, on average, we produce three times more energy from coal in this country than we did thirty-five years ago, and we have one third of the emissions, of the regulated emissions. Sulfur dioxide, nitrogen oxide, those emissions have been cut by about seventy percent overall. It's an affordable energy resource. It's about half the cost of using other fuels. It's a fuel that we've used primarily to generate electricity here in this country. Over half of our electricity comes from coal. It is an energy resource that could be looked at to do some additional things." Such as converting it to create "clean diesel" and other transportation fuels. But what really makes coal an option for the future, boasts Lucas, is the potential to clean up the coal-fired power plants so that they emit no more pollutants.

"Those power plants will have zero emissions. They will not emit any of the pollution that we're looking at today: sulfur dioxide, nitrogen oxide, even mercury. Also, because we're concerned about global climate change and reducing the carbon intensity of the energy sector both here in the United States and around the world, that will be possible as we capture CO_2 for permanent storage in underground aquifers. And these power plants will also produce hydrogen as a by-product. And that hydrogen can be used to fuel automobiles. So I'm very optimistic about the future of coal and where technology allows us to go with that."

Not so fast, literally, says Jeff Goodell, author of *Big Coal: The Dirty Secret Behind America's Energy Future.* "The coal industry is very good at touting new technology and less good at actually doing anything about it. There is new technology that's available now, called IGCC, integrated gasification combined cycle, a kind of a gasification of coal. But the industry has resisted building these plants. They prefer to tout these plants that are ten or twenty years down the road and continue building the same old thing.

"The fact is that carbon dioxide from coal plants has gone up about twenty-seven percent since 1990, and they're continuing to go up. And global warming is an increasing, very urgent problem. We need to cut emissions, most scientists agree, by fifty percent or more by the year 2050. And the coal industry is going in exactly the opposite direction. The only way that you can think about coal as clean is in just a very narrow way of thinking about it. The fact is that coal can only be considered clean by the narrowest of definitions. It's true that the levels of air pollution of sulfur dioxide and nitrogen oxide that Joe mentioned have fallen. But one of the things he doesn't mention is that the coal industry fought tooth and nail against all of those laws that required those reductions during the '70s and '80s and '90s, spent millions of dollars lobbying against them."

As for the greenhouse gas CO_2 emitted by burning coal, what about sequestration? Pumping CO_2 underground, storing where it cannot get back into the atmosphere?

Goodell certainly agrees that "sequestration will certainly work. It's working now, and they're doing it in many places in the oil industry. They pump CO_2 down underground into oil fields to help push the oil out. There's no question that we can build coal plants that can sequester CO_2 underground. The question is, first of all, how soon we will do it. The second question is, its capacity is quite limited. It works in certain places, in certain geologic areas where the structure of the geology underground is just right. For example, it

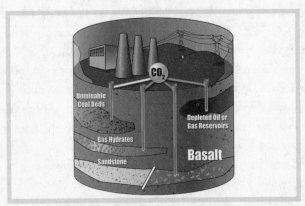

Carbon dioxide sequestration: Researchers are trying to find the best geological formations for safely storing carbon dioxide. Pacific Northwest National Laboratory.

worked great in Montana or Wyoming, but it's not going to work at all in the southeast."

Another, problem, says Goodell, is a cultural one. "We haven't even begun, as a society, to think about [this]: Do we want to embark on this huge campaign to bury millions and millions of tons of CO_2 underground, all over America? The biggest sequestration field right now is in Canada. And it will, after twenty-five years of use, sequester about twenty-five million tons of CO_2, which sounds like a lot, but a coal plant in Georgia emits that much in one year.

"So do we want to be living above these giant bubbles of CO_2? CO_2 is an asphyxiate. A natural bubble of CO_2 was released from a lake in Africa in 1987 and asphyxiated seventeen hundred people. Do we want to be living above this? We haven't even begun to think about it yet."

"I disagree," says Lucas. "We're in cooperation with the federal government, in several states. There are these sequestration partnerships that are being funded by Congress, being enacted through the Department of Energy, where lots of communities, lots of stakeholders are out right now studying the various applications of where you can store this, what it will mean to do this long term. I think we are very much addressing what the possibilities are of this for the future, as we go forward. It is clearly something a lot of progress is being made right now on this very issue."

As for the delay in building those new IGCC power plants, Lucas points out that three of them were built, one in Indiana, one in Florida, and one in Nevada. But a design defect in the Reno plant and "deregulation" of the power industry is "sort of the reason why more IGCCs were not built."

ENERGY AND POLITICS

You might think that technology is a black-and-white issue: Either you have the technology to create a new generation of coal-fired power plants or you don't. If not, you need more time to develop it.

That's hardly ever the case when you're talking energy. Energy is so entwined with politics that it's impossible to talk about one without considering the other. (Think about those secret meetings Vice President Dick Cheney convened with the major energy companies, the minutes of which the White House refuses to release. Just read the Department of Energy Web site, which says the FutureGen project "will be led by an industrial consortium representing the coal and power industries.")

For example, when you ask Lucas whether we need to create any new technologies to create this pollution-free, coal-fired FutureGen power plant, he says, "Basically we have them." But when you press and ask him, "Why, if the technology exists to build such plants, is the industry waiting ten years for a demonstration project?" he points to the typical engineering problems in scaling up a technology, from a small testing phase to the size where it can be employed. "You're making a bunch of technologies that have not worked together have to work together for the first time. Since this is a first-of-its-kind project, first-of-its-kind technology, first-of-its-scale technology, let's say they build it and it doesn't work. Do you think the utility is going to take that $100 billion loss itself?

"The other thing is—and I don't think people realize this—just the siting of a power plant takes sometimes five to six years. And so from that standpoint once that site selection takes place, it will take several years to get all of the permitting authorities that are necessary to move forward with that plant and then begin construction of that plant."

Goodell says he has heard all of this before. "It's the same story over and over. It's always, 'Oh, if we can just do a little more research, if we can just get a few more grants, if you can just give us a few more hundred million dollars, we'll figure this out.' It is not a technological problem. It is a political problem. It is a political problem because it is not a level playing field. The coal industry is one of the most—if not *the* most—politically powerful industries in the United States."

The energy industry explains delays as "uncertainty." Energy critics call it "stalling." "This kind of redirect is exactly what happens whenever there is any issue," counters Goodell. "It was the same talk in the '70s. We can't clean it up. It's too hard. It's too expensive. The stuff's not ready yet. And then they pass laws and it becomes ready and they do it far cheaper than before. It was the same debate with mercury (pollution). We can't do it. It's too hard. Stuff's not ready. You know, if we passed a law, they would do it.

"It's the same thing with IGCC. Gasification is widely deployed around the world. There have been a number of IGCC plants that are up and running all around the world. Two years ago, *POWER Magazine*, the sort of *Vanity Fair* of the power industry, said, 'Coal Gasification: Ready for Prime Time.' I mean, this is just a strategy well known in the industry called 'delay and fail.' They just want to keep doing the same thing."

COAL AND CAPS?

No one doubts the political power of the coal industry. Goodell points out that presidential candidate George W. Bush pledged to place mandatory limits on CO_2 emissions. But once elected president, Bush changed his mind, Goodell says, under pressure from coal company lobbyists.

"This is really the big, scary thing for the coal industry. Because mandatory carbon dioxide limits are going to happen. And that's going to change the price of electricity from coal plants and make them much less competitive, because coal is by far the most carbon dioxide–intensive fossil fuel. Why so many of these plants are rushing to be built is to get them permitted and going before these regulations happen so they can be grandfathered in and produce cheap power and big profits for these companies for years to come."

"If Jeff is right that we are going to have mandatory carbon restrictions here in this country, that's truly unfortunate," says Lucas. "Because we know two things about mandatory carbon emission standards. They're hugely expensive and they're ineffective. They're hugely expensive in that it will cost the United States economy hundreds of billions of dollars." As for being ineffective, Lucas sounds the same criticism voiced against America signing the Kyoto Protocol: If large, rapidly growing countries such as China and India do little to curb their own greenhouse gas emissions, why should the United States curb its own?

"These are places which have huge indigenous reserves of coal, must bring the power of electricity to their economy and to their people, and they're going to use that coal either with advanced technologies or without advanced technologies," says Lucas. "And that's why I'm saying [that] if you want to address the issue of climate change, you're actually bailing out a bathtub with a thimble with the water still running full blast."

"Let me say two things about that," counters Goodell. "First of all, I have eight-year-old twins, and when one of them makes a big mess in their bedroom, we have a simple rule. You make the mess, you clean it up. In a certain way, that's what's going on in the atmosphere here. The reason that the global warming, the carbon dioxide levels are so high is because of the industrialized West. So we, first of all, have a moral obligation just for having made this mess and caused this problem, to deal with it.

"Second of all, one of the best ways to help China—and I've been to China a number of times—move away from coal is for ourselves" to make the first moves. "We are the entrepreneurial power of the world. We are the innovators. We're the people who can solve this problem. And if we adopt new technologies, clean-energy technologies, they will adopt them in China also. The quickest way to drive improved technology there is for us to do it here."

"And you know what? People are figuring this out. There was $1.4 billion spent on venture capital into clean technology. We have venture capitalists who say that the economic opportunities in clean technology, clean energy, figuring out new ways to generate power are going to make the kind of prosperity in Silicon Valley in the '90s look like a bake sale."

CHAPTER FIFTEEN

THE WIND RUSH IS ON!

Texas, Kansas and North Dakota together could generate enough electricity by wind to power the entire United States.
—U.S. DEPARTMENT OF ENERGY

With all the talk about energy independence and global warming, the high cost of building nuclear power plants, and the problems of storing the greenhouse gas CO_2 so it doesn't leak back into the atmosphere (see, I told you it would make your hair hurt), there is, after all is said and done, a way to create all the energy we would ever need, cleanly and efficiently.

That way, that solution, is to change our society, change our economy, into one that depends on the wind and the sun to produce much of our energy. We've touched on it briefly in the other sections on producing ethanol or installing wind turbines. But all of those solutions foresaw the sun and the wind not as primary sources of electric power but rather as add-ons, adjuncts to nuclear or coal or ethanol. There are some people studying our energy problems who

Woodstock of wind: Farmers are eagerly reaping the rewards of the wind. These are just 2 of the 17 wind turbines on the 320-acre farm belonging to the Kas family east of Woodstock, Minnesota. They were the first farmers in the country to own their own commercial-scale wind turbines, producing enough electricity to power 4,300 homes. Photo: Windustry.

view wind and solar power as *the* solution. But it is a solution that would require staunch political leadership and many years of development.

WIND: GETTING OUT OF DODGE

How big is a wind turbine blade? Six-foot-six Wes Slaymaker stands next to a blade and hub on the construction site of the turbine at the University of Minnesota, Morris, which is a 1.65-megawatt Vestas turbine. It began producing power in March 2005 and supplies the equivalent of about half the university's electricity needs. Photo: Windustry.

Kermit Froetschner is a farmer in Spearville, Kansas, twenty miles from Dodge City. Froetschner has 16 spinning wind turbines on his farm north of town, producing such a spectacular view that it stops traffic. But what interests farmers and local residents more is that over the next 30 years, the local electrical utility, Kansas City Power & Light, will pay the county and the farmers almost $10 million for the use of their land. So while Froetschner pumps out electricity, the utility pumps much-needed hard currency back into the economy.

As for which would please him more, grain farming or wind farming? "I'll take the turbines," says Froetschner. His wind turbines are just a few of the 67 turbines that make up the Spearville Wind Energy Facility, a 100.5-megawatt wind energy–generating grid that went online in late 2006. The turbines generate enough electricity to power 33,000 homes, and the facility isn't even the largest system in the state. "It's a good supplement to keep the farm," he says proudly.

Kansas is not alone. States all across the country are discovering the power and profit of the wind, says Matt Steuerwalt, energy policy advisor to Governor Christine Gregoire of the state of Washington. "We're seeing a lot of the same thing that Kermit is describing for our farmers. We've got folks out in eastern Washington who are growing wheat right under the towers. It's a second cash crop for them."

While their crops grow and cattle graze among towering wind-mills, the typical farmer can expect to receive $2,000 to $5,000 per turbine per year in leasing fees. Virtually every farm in America (94 percent) depends on outside income to survive. When you take into account that a typical lease could last 25 years, no wonder farmers are waiting in line to sign up. There's even a Web site for a nonprofit group called Windustry touting its slogan to farmers and other land-owners: ". . . learn how to harvest the wind."

Being remote, there is not much opposition to the farmland wind farms. And where opposition might mount closer to the major cities, the governor of Washington State pushed through a tax change bill to allow the cities and counties to take some of the tax revenue from these projects. Money talks. "And you'd be amazed at what happens to the opposition when it becomes time to fund schools and fire dis-tricts and local hospitals with the revenue from these projects. So it's a real winner."

And wind turbines are among the cheapest-to-build and cleanest-to-operate power sources. "There's no fuel required, there's no drill-ing, there's no mining, there's no emissions, there's no hazardous waste to clean up," says Steuerwalt. "That's why it's such an attrac-tive option in many places where it can really compete."

I'LL SEE YOUR GIGAWATT AND RAISE YOU A GIGAWATT

And that fact is becoming very obvious. Wind is the fastest-growing energy source in the world. In the United States alone, wind energy has increased 900-fold since the 1980s. So many cities, states, and countries are racing to reap the wind that the pace of wind-energy production is staggering. One city is trying to outdo the other. Saying that it wants to be known as the nation's—if not the world's—biggest wind-energy producer, the state of Texas announced in May of 2006 that it will build the country's largest offshore wind facility, off the coast of Padre Island in the Gulf of Mexico. (In West Texas alone, more than 2,000 wind turbines currently dot the countryside.)

Just weeks after Great Britain boasted, in December of 2006, that the world's largest wind farm—441 turbines producing 1,000 megawatts (1 gigawatt) of power to provide a third of London's homes with electricity—would be built in the Thames estuary, there came a press release from British Columbia, topping their brother Brits: The construction of the 3,000-megawatt (3-gigawatt) Banks Island Wind Farm, the biggest in the world, is slated to begin construction in 2009. Oh, Canada!

I'm sure that by the time you read this, someone someplace else will probably be announcing an even bigger wind farm.

Experts believe that through 2020 to 2030, wind energy might make up 20 percent of our world energy supply. Denmark claims to have already reached that goal, with an extensive range of wind turbines. Germany leads the world with 19,000 megawatts, or 19 gigawatts. (The Schleswig-Holstein region of Germany has the greatest density of wind turbines per person in the world.) The rapidly moving United States has just slid into second with a generating capacity of just over 10,000 megawatts (10 gigawatts), with Spain right there in the same ballpark.

Farmer Froetschner says he'd love the electric utility to build more towers on his farm—"they put up three or four a day"—but there is no place for all that electricity to go. "We produce more electricity out there than what we use. I think they could double it."

The transmission lines, he says, are probably the factor that limits his capacity and distribution. They must be located within 5 to 10 miles of the electrical grid. "So we have to get it someplace where it's needed. I'm not an electrician, but we have power lines that need more transmission lines to make this farm expand."

THE SAUDI ARABIA OF ELECTRICITY

And there's the rub: bringing electricity where the wind blows, to places where the demand is highest. If you look at the wind charts of the United States, remote areas such as North and South Dakota have winds that blow strong enough and long enough to produce enough electricity to supply almost half the power of the entire country. According to the U.S. Department of Energy, just three states—Texas, Kansas, and North Dakota—have enough wind power potential to supply energy to the *entire* United States. What Saudi Arabia is to oil, these states are to wind. The problem is that the electric grid that would get the juice back to the most populous states where electric consumption is highest doesn't reach those remote areas.

One obvious solution is to extend those power lines out to the grid. That would cost tens of billions of dollars. But why not think bigger? As long as we're going to think about spending the big bucks to modernize and extend the rickety old power grid, why not try something new? Why not convert the electricity into hydrogen, then pipe or truck it to service stations to be pumped into electric cars or power plants running on fuel cells?

In other words, don't think of hydrogen as the energy. Think of hydrogen as the carrier of the energy—a universal storage system of electricity. And think of it not as a carrier just for electricity made by wind power but for electricity made by other alternative energy sources such as solar.

THE HYDROGEN SOCIETY

That way, says, John Turner, principal scientist at the National Renewable Energy Laboratory (NREL) in Golden, Colorado, you could not only make the windy plains your home for wind turbines but you could also make the very sunny states of the Southwest your home for solar and photovoltaic generators.

Turner believes that given all the sources of sustainable energy in the country—energy that never gets used up—we could in a matter of decades become totally energy independent in a hydrogen economy.

"I say, 'Do we have enough energy to supply all the energy needs for a future society,' say, you know, eight to ten billion people? And the answer is absolutely. There is no question."

In fact, Turner says solar cells, by themselves, could supply the world's demand for energy by the year 2050. But it would take a lot of solar cells. "It would take an array the size of Texas," he says. Nevertheless, he points out, we have that resource available for use, "and we have the technologies today that can take advantage of that. That's the whole beauty of solar. You know, from villages to buildings, you can have some large arrays in various places around the world."

But because the wind and sun are intermittent—because the sun doesn't always shine and the wind doesn't always blow—"we need an energy carrier for transportation and other things like energy storage, and that's where the hydrogen comes in." Hydrogen would store the energy. "It's a chemical carrier." And we could make extra hydrogen at times of low electric demand, such as night hours.

"We're far behind in solar cells," he points out. The price of solar cells really needs to come down. "So some breakthroughs need to be done there in terms of getting our costs down, but the technology is there."

WIND + WATER = HYDROGEN

Some small steps testing the feasibility of a hydrogen economy are slowly being taken. The U.S. Department of Energy's NREL, in partnership with Xcel Energy (the same folks who built those wind turbines in Spearville, Kansas) recently unveiled a demonstration plant at NREL in Golden, Colorado, that uses wind-generated electricity to produce and store hydrogen, right there on the spot. "Converting wind energy to hydrogen means that it doesn't matter when the wind blows, since its energy can be stored right there on-site in

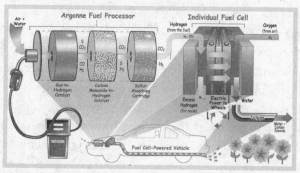

Argonne National Laboratory is researching a way to turn fossil fuels into hydrogen for a fuel cell–powered car. Here, a fuel processor turns gasoline into hydrogen plus carbon monoxide and carbon dioxide. The carbon monoxide is then converted into additional hydrogen plus sulfur and carbon dioxide, and the sulfur is removed, leaving only traces of carbon dioxide and hydrogen. The hydrogen powers each of several fuel cells, which make electricity to power the vehicle. The emissions are carbon dioxide and a small amount of water. Argonne National Laboratory image.

the form of hydrogen," said Richard C. Kelly, Xcel Energy president and CEO.

The electricity from two wind turbines is passed through water, which splits the H_2O into its components, hydrogen and oxygen. (It's the same demonstration you did in seventh-grade science class: Dip the wires from a battery into a glass of saltwater and out bubbles hydrogen from one wire and oxygen from the other.) The hydrogen is stored for use later in a fuel cell or a generator powered by an internal-combustion engine. Both Xcel Energy and NREL are chipping in to pay for the $2 million, two-year project.

It's only an experiment to test the feasibility of such a system. But if alternative energies such as solar, wind, and hydrogen are to catch on and become mainstream, "I really think it has to be a national initiative," says Dr. Amy Jaffe, associate director of the Rice Energy Program at Rice University in Houston, Texas. "There are groups of people who have called for an *Apollo*-style national initiative in science," an effort that is going to take decades, as Dr. Turner pointed out. "And so it's really important to start focusing on the science today."

For example, says Jaffe, when Turner talks about how much land the solar arrays will cover, the land itself becomes an issue, both for wind and solar. "People who are green don't often mention that. But if we can invest in revolutionary science technologies that utilize nanotechnology, whether that's to have better membranes or better solar panels or smaller this or smaller that, then I think the potential is larger."

Some scientists have already created a nanosolar solution. They have found ways of creating spray-on, plastic solar cells, made with material that uses nanotechnology. Imagine spray-painting houses or barns with nanosolar paint or rolling out large sheets of plastic solar cells to cover arid parts of the desert.

A company called Nanosolar has found a way to coat sheets and strips of thin metal with photovoltaic plastic, akin to printing ink on paper, opening up the possibility that solar panels could be placed on any building surface exposed to the sun. Nanosolar recently announced it would build the world's largest factory for producing solar cells in San Jose, California. Working at full steam, so to speak, it could turn out enough solar cells each year to produce more than 400 megawatts of electric power—three times the amount currently installed in the United States.

Producing wind power in some states and solar power in others is all fine and good. But Jaffe says we won't see any real progress "until we have a real direction," as we did in the early days of the space program, which turned all resources to getting to the moon. "You're going to spend a billion dollars a year for ten years, just on fundamental science because the kinds of technologies that are here with us today are technologies that require huge breakthroughs, especially in storage technologies."

In the space program, milestones were set and met. Technologies were developed to meet each target. The same thing needs to be done in energy, says Jaffe. "We need to set goals and targets. We need to know where we need to be in what year. We need to lay these things down together so we understand the science that's possible at which time in which fields, so we understand what fuels are going to provide us an escape from emissions and which ones aren't, and that we have a coordinated national policy."

A coordinated national energy policy is not what the United States has at the moment. And weaning a country off a coal- and oil-based economy won't be easy, she points out, because so much money and politics are tied up in these industries.

"There isn't anybody who makes a living off of the sun, and so nobody advocates for the kind of technologies that John [Turner] is talking about. But there are coal states in the United States that have great political power and there are certain states that make money from having traditional combustion engines stay on the road, so we have a problem in our political process in terms of going through the evaluation. Not just of what's technically practical, financially practical, economical and commercially practical, but when you add a layer of the politics of trying to implement what's best for the country as a whole, then it becomes much, much, much more difficult to do."

One thing that is not in doubt is the cost of electricity generated by the wind. It has declined dramatically. With the advent of larger, more efficient wind turbines, wind-generated electricity is now competitive with other industries. Depending on the site, wind-power electricity is three to seven cents per kilowatt-hour, says Laurie Jodziewicz, communications and policy specialist at the American Wind Energy Association. "In these days of high costs for natural gas, wind energy is actually bringing down the costs of electricity to some consumers by offsetting that need for more natural gas use."

It's hard to argue against wind power, though some have. Some of the opposition comes from people who fear that the turbines may kill birds and bats. "Even if we got one hundred percent of our power from wind power—which is probably not realistic—but even if we had one hundred percent of our power from wind, the bird impacts would be very minimal, compared to things like buildings, cats, vehicles, pesticides, and all of the other things that affect birds," says Jodziewicz.

"With regard to bats, there was something that was unexpectedly discovered in 2003. And the industry immediately partnered with the Fish and Wildlife Service, with the NREL, and with the leading bat organization in the world, Bat Conservation International. We formed together, and we've been funding research to understand and hopefully solve the issue that we discovered by better understanding what might make our site risky for bats but also other ways to deter bats away from wind turbines. I think that overall our environmental impacts are minimal. But we certainly want to make sure that we take care of whatever we can."

Others argue the windmills are unsightly. They'll spoil the view. It's a subjective opinion that can't really be countered, except by those who live near them and find them majestic.

But if wind power contributes to our energy independence and helps counter global warming, finding enough homes for those wind turbines will be easy, once we view them not for their size but as symbols of our security.

PART V

—

NANOTECHNOLOGY

CHAPTER SIXTEEN

THE NEW SMALL IS BIG

If I were asked for an area of science and engineering that will most likely produce the breakthroughs of tomorrow, I would point to nanoscale science and engineering.

—NEAL LANE, FORMER PRESI-
DENTIAL SCIENCE ADVISOR

The prefix *nano-* has entered the lexicon, as in "I'll be done in a nanosecond." Or the brand name iPod nano. You get the general idea that a nanosecond goes by even more quickly than, say, a New York minute. But you may not know that *nano-* simply means "billionth," from the Greek for *dwarf*. So a nanosecond is a billionth of a second, and a nanometer is a billionth of a meter, or about five to seven atoms in length. That's tiny! (And that means that the new, smaller, slimmer iPod is not "nano" in the true sense at all.)

You may have heard people mention *nano-*, but you may not have heard much about nanotech or nanotechnology. What is it? One of the architects of nanotechnology, British chemist Sir Harry Kroto, defines nanotechnology and nanoscience as "molecules that do things." Researchers in these new fields work at that incredibly small scale of

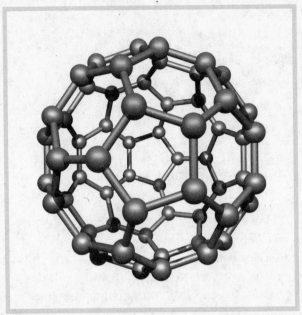

The buckyball named after the favorite shape—geodesic dome structures—of Buckminster Fuller, is a molecule consisting of 60 carbon atoms. This nanoparticle was discovered by accident but set off a revolution in physics in the 1980s. © Chris Ewels, www.ewels.info.

molecules and even individual atoms to create new materials, new processes, and new machines that could improve our lives enormously.

It all started with physicist and Nobel laureate Richard Feynman. In 1959, Feynman gave a talk at California Institute of Technology entitled "There's Plenty of Room at the Bottom," in which he challenged his fellow scientists to come up with tiny, molecule-sized machines that can do surgery, libraries that can be stored on the head of a pin (the entire 24-volume *Encyclopaedia Britannica*), minuscule computers. Why? Because small machines could work more efficiently, using a lot less power, and manufacturing them would be much cheaper. But to realize Feynman's vision, researchers needed new tools.

A big step into that very small world came in 1990, when

researchers invented a new kind of microscope, called the atomic force microscope. It has a tiny needle that bumps over atoms the way the needle in an old-fashioned phonograph jumps over the grooves in a vinyl record. This needle also can move atoms and molecules around. Scientists found that they could use the needle to manipulate and rearrange atoms—work on the nanoscale, that is—and make tiny new things. Dr. James Gimzewski is known to his friends as "Jim-Get-Me-Whiskey." One of his inventions is a "nano nose," a tiny sensor that can distinguish between different types of whiskey. Gimzewski was then a group leader at IBM's Zurich Research Laboratory in Switzerland, where he pioneered ways to manipulate atoms and molecules to make tiny sensors and machines with the atomic force microscope. Now he's a professor in the Department of Chemistry and Biochemistry at the University of California, Los Angeles, where he built his own new microscope.

In the late 1990s, the U.S. federal government began investing large sums of money in labs like Gimzewski's and other nanotech researchers'. That seed money has about doubled since. One big reason is that nanotechnology could revolutionize electronics by giving us much smaller, more powerful electrical devices that would save a great deal of energy. And we badly need an alternative to today's silicon chips. In 1965, Gordon Moore, one of the founders of Intel, predicted that the number of electronic circuits on a silicon chip would double every year—a rate that, as circuitry shrank and got more complex, he updated in 1975 to every two years. Today, it's about every 18 months. But Moore's law won't hold true much longer, because there's a limit to how small you can shrink electronics before heat from the circuitry on the chip begins to melt the plastic from which it's made. So a major goal of nanotechnologists at companies such as Hewlett-Packard, Lucent, Intel, and IBM is to shrink computer chips down to the size of a single molecule. But so far, there have been only some demonstrations done in the lab of how to build such a chip. You won't find anything available at RadioShack. Both Gimzewski and Horst Stormer, a Nobel laureate in physics who works at Lucent, say that this goal of a working chip the size of a molecule will be very hard to attain. "Right now, we are far, far from this," emphasizes Stormer.

Sandia National Laboratories' nanotechnologist Jeff Brinker is approaching the next generation of electronics another way. "I like that 1960s slogan 'Power to the people,'" he says. "I like developing technologies that anyone can use." One approach Brinker particularly likes is "smart ink," which he says that "you write with just like you do with dumb ink." Loaded into a regular printer, smart ink would allow anyone to design and print out working electronic circuits on everyday printing paper.

While some researchers are focusing on tiny transistors and circuitry, others dream of putting nanosized particles together to make much bigger things that could be incredibly useful. The late Richard Smalley, Rice University chemist, won the Nobel Prize along with his British colleague Harry Kroto for discovering the fullerene, a nanoparticle that resembles a soccer ball because it's made up of hexagonal molecules. Smalley and Kroto gave the fullerene its name and nicknamed it the buckyball because its hexagons look like those in the geodesic domes that visionary architect Buckminster Fuller unveiled in 1954. (Fuller, who was dedicated to doing more with less, would have appreciated nanotechnology.) Smalley also referred to carbon "nanotubes," tiny tube-shaped versions of the buckyball, as buckytubes. Despite their minuscule size and the fact that they're made of carbon, the same stuff that's in your pencil lead, nanotubes are incredibly strong, yet as light and flexible as straws. They're an excellent example of how very differently things work at this incredibly small scale.

To understand carbon nanotubes, one prominent nanotechnologist, Cornell University's Paul McEuen says, "Think of a stack of paper in which each paper is one atom thin, a sort of chicken-wire mesh of carbon atoms." Unlike nanoscale circuitry, carbon nanotubes are already in products you can buy: They reinforce your car's dashboard and tires, making them stronger and longer-lasting, and also go into your skis and the frame of your tennis racquet and your bike.

Besides being strong, Smalley pointed out, carbon nanotubes also conduct electricity. Smalley believed that once we figure out how to align them and make them into long cables, they could transfer energy far more efficiently—revolutionizing energy conservation. Others think that carbon nanotubes could make space travel much cheaper and easier. Arthur C. Clarke, in his 1953 sci-fi novel *The Fountains of Paradise*, describes a "space elevator." Such an elevator would have a 24,000-mile long cable, one end anchored on Earth, the other on a satellite orbiting the Earth. Just like an elevator in a skyscraper, people would ride this space elevator into Earth orbit. Carbon nanotubes may be just strong and flexible enough to serve as the elevator cable. (See more about this idea in Chapter 17.)

Meanwhile, some nanotechnologists, such as McEuen, are investigating other uses for carbon nanotubes. McEuen has made "guitar strings" out of carbon nanotubes. Each one is "clamped down at both ends," he says, "and vibrates just like a guitar string vibrates. There's the fundamental and the harmonics, just like there are with a regular guitar string." McEuen wants to use his "guitar string" to weigh and measure atoms and molecules and learn more about their chemical composition: "The heavier a molecule was, the more it would shift the frequency at which the string vibrates. So if you listen for that change in tone, you could infer the mass." He says, "The way we listen to the nanotube is much the same as the way you listen to a radio broadcast. We take a high frequency signal and we sort of convert it down to a lower frequency where it's simpler for us to hear. So you could imagine in the future using these nanotubes as a kind of simplified radio receiver, and it might be simpler and use much less power than an existing radio."

BIONANOTECHNOLOGY

Nanotubes can be grown in a whirling hot plasma gas of hydrogen, carbon, and metal particles. This computer artistry imagines how that might occur. As the "hot soup" cools, carbon condenses around the metal particles, forming carbon nanotubes. © Chris Ewels, www.ewels.info.

Some nanotechnologists are experimenting with nanowires, incredibly tiny wires that could become part of minuscule transistors and electronic circuitry because they have optical and electronic properties. At Harvard, chemist Charles Lieber has combined them with nanoscale lasers for use in photonics, the process by which silicon-chip circuitry is now made, on a tiny scale. Lieber cofounded Nanosys, a nanotech startup company with, he says, "the modest goal of revolutionizing chemical and biological sensing, computing, photonics, and information storage."

Lieber thinks that nanowires could be very useful in medicine. He says that one of his nanowires is made of silicon, and its dimensions are similar to carbon nanotubes. In other words, incredibly tiny! He says that "this very small wire acts as sort of a switch, and then when a biological molecule binds to it, it can change the resistance or conductivity of that wire, either turning it on or off. That provides us with the selectivity to recognize one virus out of a whole soup of many different biological species. The virus binds to an antibody in your blood, and by using chemistry, we have linked antibodies to the surface of the nanowire.

"We've been able to demonstrate unambiguously that when a particular virus binds to the wire's surface, the electrical signal changed. If you could detect a virus at this early stage, when your body's immune system might be still holding it in check, you could then be treated effectively before the virus began replicating rapidly and became highly infectious."

Lieber looks forward to "detecting a virus in real time," or even many viruses simultaneously. That would mean you'd visit your doctor, give a blood or saliva sample—and get your diagnosis right there. You wouldn't have to wait several days for test results to come back from the lab before you could find out why you haven't been feeling well. If you turned out to have a serious illness, such as cancer, that early diagnosis could make a huge difference in your prognosis and treatment. On-the-spot diagnosis could save your life—or the life of a soldier who's been exposed to a chemical weapon or bioweapon on the battlefield, or the health of a swamp or river at risk of pollution.

Nanotechnologists such as Lieber and Brinker have succeeded in making us safer by developing supersensitive sensors to detect anthrax or other biological or chemical warfare agents. One of these supersmart "noses" that Brinker worked on is already in public places, such as airports and public transit systems. Another kind of sensor, called "smart dust," glitters like a disco ball from the 1970s, and turns from green to red when it detects a pollutant in the environment.

NANO ALL AROUND

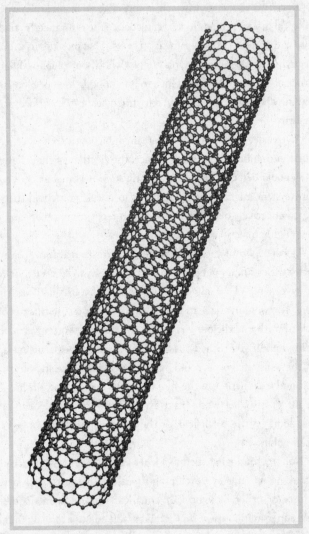

Computer simulation of carbon nanotube. Note how it is made of many rings of carbon, which when enmeshed look much like chicken wire. Nanotubes are finding their way into everything from car wax to cosmetics, solar energy to medicine. © Chris Ewels, www.ewels.info.

You may not have noticed, but nanotech has already produced some new materials that have crept into our lives. You already may be relying on nanotech throughout your day. For instance, when you begin or end your day in your bathroom, you may put on lotions, creams, makeup, and hair dye or hair gel. Some cosmetics companies are putting incredibly tiny nanoparticles in makeup, to help eye shadow and lipstick color last longer, and in face cream, to help it absorb faster and deeper into your epidermis, where it can make a difference in your skin. The world's largest cosmetics company, L'Oréal, won't say exactly which of its products are made with nanoparticles, but the company is incorporating nanotechnology into its research in a big way. One *Wall Street Journal* reporter experimented on her own face with several new antiwrinkle creams. The only one that made a visible difference in the reporter's before and after photos was a cream made by Lancôme, one of L'Oréal's more expensive brands.

L'Oréal also studied the nanoscale structure of the wings of the *Morpho* butterfly, the big beautiful one with bright blue iridescent wings. Researchers discovered that there is no blue pigment at all in the butterfly's wings; the blue you see is an optical illusion produced by the wings' molecular structure. The researchers borrowed the secrets of the wings to produce iridescent lipstick and nail polish that will be available soon.

In *Oryx and Crake*, Margaret Atwood's 2003 novel about the future, the hero moves into an apartment where the wallpaper changes color. That prediction is already coming true. Nanotechnology is shrinking electronics that you could either wear or embed into your living room's wallpaper or sofa cushions. Minuscule electric circuits that change a fabric's color via LED technology is beginning to find its way to clothing stores. There is already wall fabric that is powered to change color.

So chances are good that when you get dressed, you're putting on clothes brought to you by nanotechnology. Nanotech is already in trousers and children's clothes. Maybe you own a pair of those khakis that simply refuse to stain. If you spill root beer or red wine or even soy sauce on them, the liquid simply rolls off in little balls or droplets, like mercury—leaving your pants clean. Well, that's thanks to nanotechnology. Like running shoes and today's silicon chips, your stainproof khakis were born in a California garage. In the mid-1990s, engineer David Soane left a teaching job at the University of California, Berkeley, to devote himself to invention. He had noticed that most stain-proof cotton requires a plastic coating that makes it hot, stiff, and uncomfortable to wear. Instead of applying a coating to regular cotton, Soane worked at the nanoscale, doctoring cotton molecules to make them more like the skin of a peach. When you wash a peach, the fuzz on the skin causes water to roll right off in droplets, instead of being absorbed. That's what Soane wanted to see happen on the surface of cotton. So he added tiny hairs, an artificial equivalent of peach fuzz, to cotton molecules. When you or your child spills something on clothes made from Soane's cotton, the minute hairs prevent the liquid from absorbing into the cotton. You can't see or feel the tiny hairs; the cotton looks and feels no different from regular cotton, so it's lightweight and comfortable next to your skin.

And when you need to do laundry? Well, nanotechnologists at Samsung have come up with a washing machine that destroys odor- and illness-causing bacteria, disinfecting your clothes so that, in theory, at least, you need to wash them less often. This washing machine works by using silver. You've heard that old tag about the boy "born with a silver spoon in his mouth"? Originally, it referred to good health, not wealth. Silver ions and silver compounds can destroy some bacteria, viruses, algae, and fungi, but without the poisonous side effects of heavy metals such as lead or mercury. Before antibiotics, silver compounds were used as germicides. The new Samsung machine generates nanosized silver ions in water that latch on to and destroy any bacteria in your laundry. Then, during the rinse cycle, the machine coats your load with silver ions, which prevent any bacteria your clothes touch from reproducing.

SELF-ASSEMBLY

If you want to learn how to build something, you find a teacher who has been at it for many years, who has lots of experience. That's why scientists are turning to nature; nature has billions of years of building experience. She has built countless species of animals both large and small, hard and soft.

One of the most important areas of nanotech research is "self-assembly"—finding ways to prompt atoms and molecules to put themselves together in useful ways, just the way living cells in your body do. Some researchers are working on ways to grow back broken bones or damaged nerve cells. Others are making tiny machines.

For medicine, one of nanotech's biggest dreams is a real-life version of *Fantastic Voyage,* the classic science-fiction film in which a team of doctors shrinks themselves and travel in an equally tiny submarine through a patient's bloodstream to remove a blood clot from his brain. Some researchers are working on tiny devices that could travel to a particular organ inside a patient's body and perform repairs or deliver drugs directly to a cancerous tumor, sparing a patient from the painful side effects of chemotherapy. How is that tiny machine going to move?

LOCOMOTION, THE NANO WAY

"There's an interesting controversy in nanotechnology as to what the best strategy is," says George Whitesides, professor in the Department of Chemistry and Chemical Biology at Harvard and executive director of the National Space Society. "At the beginning, there was an idea that the way to think about nanotechnology was to look around at things like submarines or automobiles or motors, devices that work in a scale we're familiar with, and then make them very, very small, changing their size by a factor of a million or maybe more. But the idea of actually building very small mechanical motors has a lot of problems. For example, it's not clear how you power them. Friction gets to be much more important at small scales. Things just don't scale from large to small very well."

So Whitesides turned to the ultimate nanotechnologist: nature itself. "Nature's been at it for a while and has some pretty clever solutions, much, much cleverer than the ones that we can come up with right now. So why not use them? Biology is full of rotary motors, linear motors, pumps. Every time I look into the strategies that simple organisms use to sense their environment, I'm just amazed at their sophistication."

To test the concept, he used a tiny, one-celled algae as a nanoscale tugboat. He and his team used chemistry to attach plastic beads to the outside of algae, which swim by beating their flagella in a movement that resembles the breaststroke. They're also sensitive to light, so by shining light on the swimming algae, the researchers can guide them back and forth to where they want them to go.

Whitesides is also very interested in self-assembly, "the process by which some complex system puts itself together. The pieces come together of their own volition. You don't have to reach in and cause them to do it. Every crystal does that. You and I are examples of something that no robot put together; we put ourselves together. Biology is the master of self-assembly."

Nanotechnologists have succeeded in making some amazing, incredibly tiny working robots that combine artificial and living parts. One researcher uses DNA itself to make a mini-robot "walk." Since even a tiny portion of heart muscle beats, another nanotechnologist has attached a heart muscle to a nanoparticle to power a tiny working machine. But so far, these wonderful living machines remain demonstrations that work in the laboratory, not in a physician's office or a hospital—yet.

The day when medicine mimics science fiction is still quite a way off. Some predictions for nanotechnology are very unlikely to come true. For example, Smalley always pooh-poohed the notion of "molecular manufacturing"—a day some futurists foresee when all the products we need will be made inexpensively by nanobots. Nanotechnology is very difficult because researchers can't specialize—to be successful, they often have to be very well versed in several disciplines, such as engineering plus biology plus chemistry. How many people can be experts in all those fields at once?

THE DNA TRANSISTOR

Professor Uri Sivan, chief of nanotechnolgy at the Technion Israel Institute of Technology, has coaxed DNA into building transistors, the basic building blocks of all computers. "We use DNA molecules as a template to complete the assembly of this electronic device. It turns out that DNA and its related proteins can indeed build remarkable structures." The DNA is not electrically active itself; it does not carry electric current or do any of the work. "Our strategy is to use the DNA and those related proteins and engineer them in such a way that they will assemble nonbiological materials for us, and those nonbiological materials are the materials that have the electronic functionality. So we were using [the DNA] to pick up a carbon nanotube that can serve as a transistor and localize it on a DNA template. And then in the second step, we used that template to grow conductive wires connecting the nanotubes." The DNA assembles the carbon nanotubes into transistors. And then the wires are assembled, literally, on the backbone of the DNA. "We developed a number of years ago a metallization process by which we can coat DNA specifically, so metal grows only on DNA. Metal grows along the DNA template and contacts the nanotube."

Sivan compares it to taking the blueprint, the DNA instructions for making a transistor, putting it into a test tube along with all the necessary electronic ingredients to build the transistors, shaking it up, and watching the finished product come out. The transistors assemble themselves. "The last step is [that] we just dip a silicon wafer in solution. We pull it out and we have billions of those transistors on the silicon wafer." They act just as conventional transistors do.

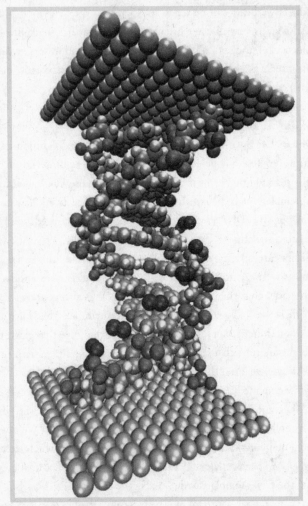

An engineered DNA strand between metal atom contacts could function as a molecular electronics device. Such molecules and nanostructures are expected to revolutionize electronics. Courtesy of NASA.

"Our challenge is to go on to more and more complex structures. The crux of the matter is whether we'll be able to invent ways to self-assemble large amounts of objects into something functional."

VIRUSES TOO

While Sivan is working with DNA, Dr. Angela Belcher, professor of materials science and engineering and biological engineering at MIT, believes that she might be able to force nature to work with materials not normally found in the wild and convince organisms, such as viruses, to build devices that are foreign to them. She is working to manipulate "three different parts of a virus simultaneously to start building" the basic parts of a transistor. "What we're doing is mimicking nature, like how an abalone grows calcium carbonate to grow a shell. We're using viruses to grow any kind of material we're interested in, and one kind of material [is] semiconductor materials."

It's a tedious process that starts with sifting through countless numbers of viruses, weeding out only those that have the right talent for the job of, say, building a wire. "We take a billion viruses and allow them to interact with the semiconductor material, and then only keep a couple of them that interact very well, and we throw the rest of them away. And any one that does interact, we keep evolving it to have better and better interaction. Then we can make billions and billions of viruses that can now grow that particular semiconductor wire. We breed them to make a wire that we're interested in."

So to create a wire, the virus would bridge two different metals. "We can actually manipulate many different proteins and many different genes on a single virus, so we can have one end of a virus grow one kind of material and another end of a virus grow or attach to another kind of material, all through this genetic selection and amplification. We pour in precursors and grow wire—it can be a metal wire or a semiconductor wire or a magnetic wire—just by throwing in precursor salts.

What happens to the virus once the wire is completed? "We can either burn off the virus at that point, just by increasing the temperature, or we can keep the virus around and recycle it and use it again."

Why use viruses? Belcher says she can use many different

organisms—she also uses yeast to grow materials—but she focuses on viruses because they "have this nice shape. They're very long and thin."

Letting nature self-assemble devices results in electronic devices with fewer errors, since the wires can form only in the right places. It also allows you to customize the exact dimensions of what you are trying to make, by manipulating the genetic code of the virus. You want it this long? This thick? This wide? How about this strong? Just jiggle the genetic code. As Belcher told *OpenDOOR*, the MIT alumni magazine, "My dream is to have a material that's genetically controllable and genetically tunable. I'd like to have a DNA sequence that codes for the production of any kind of material you want. You want a solar cell, here's the DNA sequence for it. You want a battery, here's the DNA sequence for it."

NOT TO FORGET BACTERIA

Not only can viruses be coaxed into building microscopic electrical parts; so can bacteria. The press release said it all: "A microbiologist discovers our planet is hardwired with electricity-producing bacteria." In other words, scientists have found that under certain conditions, some common bacteria can sprout nanowires that conduct electricity. And with the Earth populated with more microbes than any other form of life, that's a lot of nanonetworking.

"Earth appears to be hardwired," is how Yuri Gorby, staff scientist at the U.S. Department of Energy's Pacific Northwest National Laboratory put it. Gorby and his colleagues discovered that they could coax some microbes to transform toxic metals into sprouting microwires, called pilli, as small as 10 nanometers in diameter. These wires could be formed into bundles as wide as 150 nanometers. And many other bacteria, not in the toxic metals business, can also form these wires, such as microbes involved in photosynthesis and fermentation. But what they all have in common is the ability to reach out and touch other bacteria by growing these wires from their cell membranes that find the other microbes and "form an electrically integrated community," says Gorby.

Being electrically conductive means that the bacteria hold the potential (pun intended) to be the power sources for fuel cells and bacteria-powered batteries.

Why would nature make such bacteria that can produce and conduct electricity and have the power to clean up toxic metals? Gorby can only speculate.

"The effect is suggestive of a highly organized form of energy distribution among members of the oldest and most sustainable life forms on the planet."

SAFETY

One question often asked is "How safe is nanotechnology?" What if all those tiny nanoparticles spill into our water supply? Or what if we were to breathe them in as they came out of an aerosol spray can or sprayer in the workplace? Right now, we don't know, although we're sure that self-replicating nanobots will never take over the Earth and reduce every living thing to "gray goo," as computer scientist-entrepreneur Bill Joy has warned. Nor will we ever be stalked by intelligent, predatory nanobots, as hapless scientists were in Michael Crichton's scary thriller, *Prey*. But many people, including some in the federal government, are concerned about the potential health hazards posed by these tiny particles that can easily get into the blood and be captured by the lungs.

"Major efforts are underway in both industry and government to realize the amazing promise of this technology. However, very little attention is devoted to assessment of health risks to humans or to the ecosystem," says the National Institute of Environmental Health Sciences, part of the National Institutes of Health, in a 2003 report. "The toxicology of nanoparticles is poorly understood, as there is no regulatory requirement to test nanoparticles for health, safety, and environmental impacts. More research is urgently needed, as there are many indications that ultrafine particles could pose a human health hazard. Research is now showing that when harmless bulk materials are made into ultrafine particles, they tend to become toxic. Generally, the smaller the particles, the more reactive and toxic are their effects."

An animal study, reported in 2004 by the Society of Toxicology, compared the effects of different, common pollutants on the lungs of lab rats. It found that "if carbon nanotubes reach the lungs, they are much more toxic than carbon black and can be more toxic than quartz, which is considered a serious occupational health hazard in chronic inhalation exposures." (Scientists were from NASA's Johnson Space Center, Wyle Laboratories in Houston, and the University of Texas Medical School at Houston.)

"I don't know that we can say that nanoparticles are inherently risky," says Dr. Kristen Kulinowski, executive director of education and public policy at the Center for Biological and Environmental Nanotechnology and director of the International Council on Nanotechnology at Rice University in Houston, Texas. "What I would say is that the size and surface chemistry of nanoparticles raises concerns that they might have unique toxicological profiles that we don't see in particles that are larger of the same chemical composition."

In April 2006, the RAND Corporation released a report exploring health risks associated with the use of nanomaterials in the workplace. The report said that the U.S. government is providing insufficient funding to understand and manage risks that nanomaterials pose to the health of workers in the rapidly growing nanotechnology industry. The RAND report said that the government has directed more than a billion dollars annually to the development of nanotechnology, but just 1 percent of that, $10 million, to studying research understanding and managing the risks involved. Basically, that says there's not enough money going into researching the health effects of nanotechnology. This is an issue that bears close watching.

PART VI

—

LEAVING THE EARTH

CHAPTER SEVENTEEN

THERE'S NO BUSINESS LIKE SPACE BUSINESS

*Our goal is to solve, or help solve, what I consider to be, by far
and away, the great problem of space, which is the cost of getting
there.*

—ELON MUSK

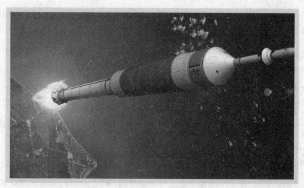

Artist's rendition of the planned Orion crew exploration vehicle aimed at returning humans to the moon. Notice the escape rocket attached to the nose of
the ship, which would whisk the astronauts away from a dangerous failed
launch. NASA/John Frassanito and Associates.

In 2004, President George W. Bush announced a new goal for the
U.S. space program: return to the moon, and after that, aim for Mars.
NASA administrator Michael Griffin said that the current space

shuttle program and International Space Station (ISS) were "not the right path" for the space agency after the highly successful *Apollo* moon missions of the '70s.

So after costing billions of dollars and two deadly accidents, the space shuttle will be put to rest by 2010. Its successor, the *Orion* Crew Exploration Vehicle dubbed *Apollo* on steroids, is now in the planning stages. It appears that we're picking up where we left off in 1975—we're going back to the future.

"We're now in a very uncertain transition that NASA and the country hasn't faced in decades, trying to design a new space vehicle while recovering from an accident with the old one, at the same time trying to draw plans that will credibly get us out to places we've never touched before," says Tom Jones, a planetary scientist and a NASA astronaut from 1990 to 2001. He flew four shuttle missions and led three space walks during the construction of the ISS.

The two shuttle disasters have convinced NASA that the shuttle's design had been faulty from the very beginning. "The shuttle was built in the early '70s, designed in the early '70s as part of a compromise in cost and capability with the government and its budgeters, and so its very design was made fragile and vulnerable three decades ago. And now we're still paying the price for that."

We were told back then that the shuttle would be so safe that even civilians would be welcomed aboard, journalists and schoolteachers. The space shuttle was the first rocket ship in the history of the space program—American or Soviet—that did not provide an adequate means of escaping a disaster during launch like the kind that killed the *Challenger* astronauts on January 28, 1986. In that tragedy, the vehicle broke up 73 seconds into the launch.

"It had been envisioned that it would be a very safe, almost indestructible or infallible vehicle, and of course, our confidence in conquering the hazards of spaceflight was oversold, and so now the shuttle, as we realize, is a somewhat fragile and now aging vehicle that needs to be replaced," says Jones. "We want the crew of this new vehicle to have a much better chance of escaping any problem with their launch vehicle and coming back to the Earth with an almost bulletproof heat shield and recovery system, and that would have gone a long way toward preventing the *Columbia* tragedy from ever happening."

So NASA engineers are bringing back the tried-and-true design of the early days of the 1960s space race: The crew capsule sits atop the rocket, safely away from the lethal dangers of falling chunks of foam or ice, and an escape rocket is positioned above the capsule, ready to whisk it away from danger during a launch gone wrong.

But why wait until 2010 to scrap the shuttle? Why not just go ahead with the Crew Exploration Vehicle now? Because according to Whitesides, the space shuttle's mission has not been completed: "We need to finish the International Space Station and fulfill our commitments to the international partners who have been literally working for a decade or two on their pieces of the space station. There's a lot of debate about the ultimate value of the International Space Station right now, but it is a triumph from the perspective of international cooperation, and we owe it to our international partners to fulfill the promise that we made, which was, essentially, to launch their laboratories and their modules."

Jones agrees: "The vehicle has a job to do in the next few years to get the most important components that are on the ground, already checked out, up to the space station. Those include the Japanese and European laboratory, for example, that our partners have been waiting on for years."

But some experts feel that finishing the space station is throwing good money after bad.

"It's very ironic. NASA already is planning to walk away from the space station as soon as they get it built," says Rick Tumlinson, founder of the Space Frontier Foundation. "Not only are we draining money to keep the shuttle going, which I think is partially political, [but] there's a large standing army of political constituency, people who depend on shuttle-related jobs to keep going."

Tumlinson says he understands those promises to foreign countries, but there could be ways around the commitments. "We have created and designated a national space transportation entity in NASA, and come hell or high water, they're going [to] hang on to their own ability to have access to space, which I think is not what we need to open the frontier. Whereas what I would rather see is all kinds of vehicles accessing space so that we can start creating a frontier up there and an economy that would allow us to expand and stay."

SPACE ENTREPRENEURS

Enter the space entrepreneurs, people of means and money, billionaires who see the "final frontier" as a business opportunity. They believe that the future of space travel belongs to travel agents who will be booking extraterrestrial trips to hotels orbiting hundreds of miles above Earth or planted firmly on the gray soils of the moon. And they already have begun to lay the groundwork for that day.

Burt Rutan and his company Scaled Composites raised the bar on private-enterprise space travel by winning the $10 million Ansari X PRIZE. It became the first company to send a person into

Paul Allen, Mike Melvill, and Burt Rutan after the first private space flight with a crew aboard. Courtesy of Scaled Composites, LLC.

space twice within a two-week period aboard a privately owned spaceship: *SpaceShipOne*. Billionaire Paul Allen, cofounder of Microsoft, financed the feat. Rutan immediately partnered with Sir Richard Branson, founder of Virgin Atlantic Airways (part of the Virgin Group of Companies), to create the space tourist company of the future. Their goal: to create a fleet of spacecraft that would carry tourists into suborbital space rides. Branson has already begun signing up thrill-seeking customers via his Virgin Galactic tourist venture.

"He plans to use derivatives of *SpaceShipOne*, carrying four or five people at a time up to the edge of space at about $200,000 per crack," notes Dr. Jerry Grey, director of aerospace policy at the American Institute of Aeronautics and Astronautics. "Now he has to get FAA approval, in addition to what Rutan got, in order to carry passengers. And that process, I think, will be fairly easy to do. Again, people will probably have to sign waivers at this stage of the game. But ultimately, I think we'll see more and more people. Branson does not invest money in things that he doesn't think he will

SpaceShipOne undergoing preflight inspection before its historic space flight. The registration number designates its goal: 328,000-foot altitude, or 100 kilometers (62 miles). Courtesy of Scaled Composites, LLC.

make money on, and I suspect that he will sell a large number of these flights."

Robert Bigelow, another multimillionaire, has created his own space station company to design an orbiting platform for commercial use. He too is offering a prize, a big one: $50 million dollars to the first person or company that can build a spaceship that can orbit the Earth twice, with a crew of five, and be able to dock with his space station.

"That is a much larger step forward, I think, than what Rutan has done," says Grey.

Other entrepreneurs are seeking to go even farther out into space. Among them is Elon Musk, CEO and chief designer of Space Exploration Technologies, otherwise known as SpaceX.

"SpaceX is a rocket company. Our goal is to solve, or help solve, what I consider to be, by far and away, the great problem of space, which is the cost of getting there.

"We're starting off with a small launch vehicle called the *Falcon 1*, named after the *Millennium Falcon*. I hope George Lucas doesn't sue us."

Falcon 1 will be a proof of concept.

"It is designed to put small satellites into orbit and test out the key technologies necessary to go bigger and to build manned rockets and manned capsules. But looking into the future, the big development at SpaceX is something called the *Falcon 9*, which is considerably larger. In its largest version, the *Falcon 9* would be capable of putting twenty-five tons into Earth's orbit. And it's also capable of missions to geosynchronous orbit and to escape.

"And the *Falcon 9* is also being [built] from the ground up with what's called a manned safety rating, so that the margins of safety in the design of components are higher than they are in an unmanned rocket. That's something that isn't all that hard to do if you design the rocket from the beginning to do that, but it's very difficult to retrofit a rocket to be man-rated."

A ONE-WAY TICKET TO MARS

Gullies on Mars, showing evidence of the past watery life of the red planet. Photo taken by the HiRISE camera aboard the Mars Reconnaissance Orbiter *circling Mars. Image courtesy of NASA/JPL/University of Arizona.*

The ultimate goal of SpaceX is not tourism, as it is for Branson's company, but a much bolder goal: the colonization of other planets.

"SpaceX is really to help enable humanity become a space-faring civilization and one day, a multiplanet species. And that's why I started the company, realizing of course that that's an incredibly difficult journey with numerous pitfalls, and perhaps the chances of us getting all the way there are very low. But that is the aspiration, if I was to describe the holy grail objective."

To do so, he must first lower the cost and improve the reliability of the service "to the point where if you want to move to Mars, you should be able to do so if you can afford the median house in California.

"It sounds a little odd to contemplate, but I think there actually is a business model, potentially, if you can make it cost somewhere around a few million dollars to move to Mars and become one of the founding people of a new planet."

Tumlinson is concerned with a business model too: the one driving NASA. The space agency has been traditionally seen as a pioneer, a trailblazer, the Lewis and Clark of space travel, going where no human has ever gone before. But space exploration has now been going on for more than 50 years, and it's time for private businesses to get a firm foothold, he reasons. After all, it was way before 50 years after the Wright brothers flew their crude airplanes at Kitty Hawk, in 1903, that private airliners were ferrying passengers. It's time to turn space travel over to the private sector, where it can be commercialized. Tumlinson believes that NASA understands his point of view and agrees with it. But he is concerned that NASA's coziness with the contractors who have been working with the agency for decades, who have received and continue to receive billions of dollars to design and build future space vehicles, will shut out the fledgling competitors.

"For example, NASA is going to be subsidizing this new rocket, which is probably going to be built by one of the two or three major contractors out there, who are already incorporating in their plans, sort of a backup plan. That they're going to use the money that they get to actually do those things Tom was talking about, which the current administrator is saying is for the private sector, which is carrying goods and cargo to and from space stations.

"So we already have the aerospace companies, the traditional aerospace companies, as I call them, saying that they're building a variation of their vehicle which will actually compete with these nonsubsidized private companies that are going to be allegedly carrying stuff to and from stations. So I see a few years down the road, we could have real trouble there, as the smaller guys get kind of booted out of the way by these heavily subsidized aerospace companies.

"If we're going to settle and stay and create the communities that Elon was speaking of, we have to make all of our decisions with that in mind. And that gets to the Lewis-and-Clark function for government and the enabling and protecting and nurturing the new private-sector companies that are going to be the ones who actually create the economy that allows us to stay. One of the big challenges is going to be to get NASA focused on a supportive role for this industry rather than the not-invented-here, do-it-yourself approach. Because as we say in our group, nobody stays until somebody pays. And I really don't want it to be the taxpayers."

Whitesides agrees. "The private sector is absolutely critical, is indeed indispensable to NASA's exploration plans. It is literally impossible for NASA to do what it wants without bringing in strong involvement of the private sector."

And a good place to start involving and encouraging the privatization of space is in the completed space station, says Jones. "So that's the first thing, to get the space shuttle's expenses put to bed. And then that'll free up, I hope, some money from some commercialization activity that will support the space station. All the supplies and cargo shouldn't be flown up on NASA rockets; they should be flown up on commercial enterprises that can bid for that business and get NASA out of the freight business and get it back in the exploring business. That may be the surface of the moon where we go or the asteroids, which I would like to see visited pretty quickly, and then, eventually, to Mars."

"What we really need is a partnership, as we move out, between the government playing the Lewis-and-Clark role on the leading edge, and this is where we screw up and have screwed up in the past," says Tumlinson. "Built into that Lewis-and-Clark function should be a constant handoff of operational activities to the private sectors, such as Elon's company, and the things Rutan might build, and others out there. Just constantly shedding. So we have what I call a lean, mean exploration machine, in the form of NASA moving outward. And then the settlers and shopkeepers bringing up the rear and creating an economy."

"All the while, filling in that commercial exploring," says Jones, "will be the commercial spacecraft that Elon talked about that can take some of the burden off the taxpayers and put it where the biggest bang for the buck is."

As for President George W. Bush's goal of returning to the moon and then going to Mars? Why not bring in private space companies to do the work, more cheaply and better than NASA could? And in doing so, have the moon project spur the growth of private industry. "In other words," says Tumlinson, drawing a complete picture, "let's stop NASA from doing any launches, human or otherwise, from Earth to low Earth orbit, and then from low Earth orbit to the moon. Let them focus their energy and attention to those types of transportation systems that would be a huge spur to this industry. It would be an incredible kick for these guys to be able to carry payloads to and from the space station, the pieces that we might need for building bases on the moon and Mars or communities out there instead of some sort of massive government space shuttle like *Saturn V*," the rocket booster that took astronauts to the moon but went no farther after the costly crewed moon program was killed to save money. "As we saw with the *Saturn V*, when that doorway closes, you're locked out of the solar system."

RESERVING YOUR SPACE TOURISM SPOT

Until NASA assigns the heavy-lifting jobs to small space companies, many will be turning to space tourism. We've had five space tourists so far, including the first woman space tourist, Iranian-American telecom billionaire Anousheh Ansari, who all paid $20 million for a stay on the ISS. All five space tourists booked their flights via a company called Space Adventures (www.spaceadventures.com), a Virginia-based firm that bills itself as "the only private company to have successfully flown clients to the International Space Station, and is the only private space exploration company with an established track record of success." Space Adventures partners with the Russian Federal Space Agency in arranging travel to and from the ISS as well as providing accommodations aboard the ISS. It plans to offer future space

Cassini Hugens mission, January 2007. 764,000 miles from Saturn: This mosaic of 36 images shows the shadow of the planet stretching across the rings. NASA/JPL/Space Science Institute.

tourists the opportunity of taking a space walk outside the orbiting station.

But the whole question of going into space on a large scale has yet to be established as a commercial enterprise, says Grey. "Somewhere between people flying high altitudes in Russian jet airplanes, which we do today, and the *SpaceShipOne* flight, which is a short flight into space and real orbital flight, that's where the dollars and cents are going to have to come out. People have to have a market to do that sort of thing. I think there will be one, but it's a question of how soon."

Some analysts predict by the year 2021, as many as 10,000 to 15,000 passengers per year will ride into space aboard suborbital spacecraft, at a cost of $700 million. Another 60 more adventurous and wealthier space tourists might be climbing aboard for orbital flights, bringing in another $300 million. Even China wants to get into the business. Having successfully put "tyconauts" in space in 2003 and 2005, the Chinese are planning missions to the moon and plan to eventually enter the space tourism business.

Bob Halterman, former executive director of the space travel and tourism division of the Space Transportation Association, has been studying the legal and technical barriers to space tourism and says the study outlined a plan to go about taking down those barriers. "We've seen a lot of progress made, both in the regulatory and the legislative," he says, "and also in the finance and in the insurance worlds."

"I think Rutan has taken what we might call the first baby step," says Grey, "but it's opened a Pandora's box of opportunity, and that's really the big thing he's done. He hasn't built a spaceship that's going to go to the moon or even go into orbit; he's built something that says, 'Hey, it's possible; we can go further.' And that's a very important consideration.

"There's a fellow named Bob Bigelow building a space hotel in Las Vegas, which will be an orbiting commercial facility," says Grey. "And it's a hotel and a lab. At some point in the future, hopefully not that far away, maybe before 2020, we're going to see vehicles like

Elon's carrying people and payloads to that private facility. At which point, if the NASA human exploration goes away, that would be sad; we would lose the Lewis-and-Clark function. But the breakout, the human breakout of the planet, would have moved to a sustainable level and we would be on our way."

Once people whet their appetite for space travel by a stay in an orbiting hotel, Whitesides believes they, like all tourists, will want to go even farther: "People will be out of the Earth's atmosphere and be able to look at the moon real clearly, and once they see that, I think they would have a desire to go. So beyond the orbiting resort hotel would be a cruise ship taking people on excursions to the moon and perhaps down to the surface and back, giving them a chance to do some prospecting and perhaps look at Tranquility Base or visit the international research station that might be there by then."

Grey agrees: "This has been progressive. First people went to the space camp at Huntsville, Alabama. And then they flew in Russian jet airplanes to the edge of space. These are commercial. Then we've got *SpaceShipOne* and Branson's suborbital flights. The next thing we're going to find, as we build that capability for orbital flights on a regular basis, not for twenty million bucks but for maybe a fair amount of money," is the desire to travel farther and farther out into space and "then eventually perhaps to the planets."

THE TANG EFFECT

The space race of the 1960s spun off new products and materials, from Velcro fasteners to Tang, the orange-flavored breakfast drink. "Space age" plastics, electronics, or fabrics were frequently cited as the unexpected but welcome by-products of space technology. Could the fertile and freethinking minds of space entrepreneurs enhance the future of science and technology too?

Tumlinson thinks so: "There are hundreds, if not thousands, of university students who have been wanting to have access to the microgravity that you can get out of a *SpaceShipOne*-type vehicle and haven't been able to do so either [because of] cost or because they're in the queue waiting for government vehicles. So that's going to open up a whole new realm. And also in the zero-G planes that are being flown out there. You're going to be able to have a lot more access to science.

"Also, in the bigger picture: The people that came to America, this new world, originally were coming for, shall we say, selfish, greedy reasons and enterprise reasons to make money. And you know, maybe that's not clean in the world of science. But in the end, because of the access they had, they created a new nation called America, and this country has revolutionized science across the board completely. So by creating access to the new world of space, where people who couldn't have afforded to go there before can get out there, we're going to see an incredible amount of revolutionary stuff happen in science, across the board."

THE SPACE ELEVATOR

That kind of out-of-the-box, creative thinking is already evident in an idea that has gained quite a few followers. It's called the space elevator, and if you search for it on the Internet, you'll find millions of references.

"It's a really hot topic," says a very excited Whitesides, whose organization is carefully studying the idea. "We know that as a society or civilization, we're serious about space exploration when we build the space elevator. It's really the Brooklyn Bridge to space."

The space elevator is a very simple idea first suggested by one of the world's great futurists, science-fiction writer Arthur C. Clarke. In his 1953 novel *The Fountains of Paradise*, Clarke's hero wants to build a space elevator. Such an elevator would have a 24,000-mile-long cable, one end anchored on Earth, the other on a satellite orbiting the Earth. Overcoming gravity—that is, getting into Earth orbit—is the most difficult and energy-consuming part of any trip into space. The space elevator makes that job much easier. You settle into geosynchronous orbit—up where those communications satellites are parked—

and drop a long but very strong wire or ribbon back down to Earth.

"Then you have an elevator, a sort of a climber that hooks on the side of it and it climbs up into space. And the wonderful thing about that is that you don't have to use rockets to get out of the so-called gravity well of Earth." You can slowly ascend and then get up to the top, and then off you go into the solar system.

So much for the theory. Actually building one is another story. The wire ribbon would have to stretch for 30,000 to 60,000 miles. What material could be strong enough to stretch that far without breaking? Hi-tech carbon nanotubes have been proposed. (See the "Nanotechnology" section for a fuller description.) Some space enthusiasts think that carbon nanotubes could make a cable for a real space elevator. NASA has invested in some studies of exactly how useful carbon nanotubes might be in making a space elevator cable and has even held a scientific conference on the subject after the *Columbia* space shuttle disaster in 2002. But so far, progress on stringing carbon nanotubes together has been slow. Carbon nanotubes are very strong, but they are also very slippery and difficult to align to make a thread, much less a miles-long cable. So right now, it's extremely expensive to make carbon-nanotube fiber.

"We've never really built anything that long ever before and not even made something as big as a pencil with carbon nanotubes," warns Musk. "I think there would be some hesitation for people to have this gigantic cable sort of spinning out from Earth and potentially becoming a hazard if it ever does break or come loose or something like that."

Whitesides agrees. The space elevator is a concept for the future. But as a metaphor, it's a dream that fits. "The reason there was a bridge built across the river in New York was the large, thriving economies on either side. By the time we have large cities and economies on the moon, Mars, on the asteroids in space between worlds, then the demand might be there for that sort of technology. And commensurately, that technology will have developed to a level to be able to fulfill that demand.

"Just as, when they first landed in Manhattan, the idea of a bridge that could span that river, of course, was something that was difficult. I think that when we begin to see these large populations moving off [into] space, that demand will grow."

THE NEXT SMALL STEPS

Grey, Tumlinson, and Halterman believe that the space entrepreneurs have taken the very first but important steps into privatizing and commercializing space. Once people take those first trips into orbit, says Halterman, they'll want more. "People have expressed their desire in surveys for a place to stay in space, so that gives credence to putting up a habitat of some sort—a resort hotel, perhaps. And then people will be out of the Earth's atmosphere and be able to look at the moon real clearly, and once they see that I think they would have a desire to go."

Grey points out that if you want to look at a model of how the government can create, stimulate, and then pass off a technology to private industry, look no further than the commercial satellite business, an industry generating about $100 billion in revenue per year.

"How did that get started? It got started because NASA invested about a billion dollars' worth of taxpayers' money over a ten-year period to create the technology. It was then turned over completely to private industry. NASA does no advanced communication research anymore. It's all done by private industry. Now see if we can use that model in transportation. NASA still has some fairly good technology programs. If NASA is willing to open this technology up for private enterprise to use, then private enterprise can pick that up and take these next steps, move a little further into orbital flight and so on. But we're not going to do it with *SpaceShipOne* technology."

PART VII

—

THE OCEANS ARE
IN TROUBLE

CHAPTER EIGHTEEN

SYLVIA EARLE: SOUNDING THE ALARM

I want to get out into the water. I want to see fish, real fish, not fish in a laboratory.

—SYLVIA EARLE

The oceans are in trouble. Ninety percent of large fish species worldwide—including tuna, swordfish, and marlin—have been decimated by over-fishing and destructive fishing practices such as deep-sea bottom trolling.

Purse seine tuna boat: The large net scrapes anything and everything it can snag on the bottom of the ocean, scouring and uprooting all wildlife. Wolcott Henry 2005/ Marine Photobank.

Seventy-five percent of all commercial fisheries have been fished to capacity and are approaching collapse, if they haven't collapsed already. Coral reefs and deep-sea habitats are being destroyed, threatened by human activities. Global warming is heating the oceans, causing them to be less hospitable to coral reefs. And in some places, coral reefs have been dying in such huge quantities that some have been almost wiped out.

Less than 1 percent of the Earth's oceans are under some kind of protection. We have protection for trees, owls, and fish in our rivers. But only 1 percent of the Earth's oceans are protected.

Just how bad is the situation? According to research published in the journal *Science*, by 2048, all fish and seafood species are projected to collapse because of the loss of biodiversity in the oceans.

Sylvia Earle is working to change that. Earle is an icon of ocean exploration. More than anyone since Jacques Cousteau, Earle, who has the biologist credentials Cousteau never earned, is perhaps the world's most visible—and vocal—marine biologist. As executive director for marine programs at Conservation International in Washington, Earle says that new fishing methods developed since the 1950s to locate and catch fish have decimated fish populations all over the world.

"The oceans of the world are in trouble as a consequence of our seemingly insatiable appetite for what we take out of the sea. And it has changed over time. Certainly the fishing issues are global. They're not limited to any one country or any one continent. And the solutions have to be global too."

AMBER WAVES OF FISH?

One of the solutions is very simple, says Earle. "Think of fish as wildlife," she espouses. "We don't try to feed six billion people with wildlife anymore. We never did try to feed that many people. But going back ten thousand years, as hunter-gatherers, we largely made a living from wildlife. It's only when we started to cultivate that we began to get to an era where we could support larger numbers, and today, about half the calories that feed the people of the world come from a handful of grasses, corn, rice, wheat. Wildlife from the sea is still contributing to the protein that we extract. But we have really not only taken the fish themselves but we've [also] destroyed much of the resilience of the ocean through the by-catch that goes along with taking the fish that are consumed, and the habitat destruction. It's really a growing problem."

Our restaurants and supermarkets have been typically filled with food fish—such as tuna and swordfish—which are very large and grow rapidly. So when we hunt and capture them from the sea, they feed a lot of people and reproduce and grow quickly. But overfishing—and our taste for sushi—has depleted stocks of those fish, leaving us with others that do not grow back so quickly. Take the newly popular orange roughy. That fish on your plate may take 30 years to mature and may actually be 100 years old! Imagine how long it will take to replace that entrée!

"So in time," says Earle, "we will necessarily have to look to ways to cultivate what we take from aquatic systems for large amounts of protein, if we are to take aquatic creatures at all."

We arrived at this crossroads because our culture teaches us that the oceans are boundless, that they are so vast and plentiful that they could never be depleted, says Earle. We believed many years ago that our forests and topsoils and watersheds were plentiful too, until the day came when they began to disappear. And we did something about protecting them.

"The [National] Park Service came into being to protect areas of natural beauty, our cultural, our historic, our natural heritage, and today, not just in North America but around the world. About twelve percent of the land enjoys some form of protection. In the ocean, it's a tiny fraction of one percent, and that's probably because we still have this attitude that the ocean is infinite and able to rebound no matter what we do. But in the ocean, we're learning not only that it can be influenced by what we do but [also that] we can take positive actions to protect what's there."

Earle says we need to find these critical areas, these hot spots in the oceans, that are most vulnerable and try to protect them. Some of these places are the sea mounts, the tops of underwater mountains, extensive mountain ranges that run down the Atlantic, the Pacific, and Indian Oceans like giant backbones. They rise from the ocean floor a thousand meters or more, yet remain below the surface.

Though hidden from view, "they're crowned with life. One of the special things about them is the apparent high degree of specific endemism, creatures that live there and nowhere else, not only in the ocean or the world but in the universe. We look at the Galapagos as a special place because it has creatures that occur there and nowhere else that we know about. The same appears to be true of these undersea islands. They don't break the surface, but they're islands nonetheless, separated from other places where such creatures can grow by many miles, and so they've developed their own character."

They are teeming with a rich diversity of life. And it's this biodiversity that makes the sea mounts very important, able to resist and recover from change. "Diversity has a really important place in the world. In terms of providing stability, when changes occur, if there's great diversity, somebody's going to be able to respond favorably to those changes. If you've got everybody basically responding in the same way, then the consequences can be really drastic."

Yet even these sea mounts, located beyond any territorial waters, or economic zones, are under attack by forces that are destroying them. Earle likens them to "the Wild West—there's little to inhibit people from doing just about anything they want to do out there." A handful of nations, she says, are "disproportionately destroying something that we don't know how to put back together again." By that she means trawling, using heavy nets that scrape the sea floor, "like using a bulldozer to catch songbirds and squirrels, they're knocking down ancient corals, sponges—the whole cross-section of life that is there, and it is simply thrown away, in order to capture a relatively small number of fish that feed not a lot of people but a high-end luxury market, people who are not really dependent on the sea for food but rather are willing to pay a high price. And all of us pay a high price because of the destruction that is brought about because of this action."

Earle is hoping that the United Nations may take action, as it did in 1992, putting high seas drift nets off-limits in some areas "because they were found to be so destructive and disproportionately favoring just a few industries and a few countries at a loss to all of us. Now that people are moving deeper and further offshore to get the last remaining pockets of protein in the sea, there is a similar concern that the cost of doing this, the cost to all of us for all time for the benefit of a few in the short term is just not worth it."

The ocean is under attack in other ways, and a cause for great concern, she says. The increase in CO_2 in the atmosphere is setting off a process that is making the ocean more acidic, a thought that is setting off alarms in the minds of marine biologists. Because a more acidic ocean means that corals may not get the calcium from seawater they need to build their homes. Nor may the other tiny, planktonlike creatures in the ocean be able to build the stony calcium-carbonate shells that are their homes.

"Now, the ocean has not gone acid yet, for heaven's sake. It's still pretty solidly on the basic side of things, but there is a trend, and just as we worry about looking at the decline of fish over the last hundred years, especially the last fifty, we're seeing a trend now in the ocean that we ought to know about at least and have on the balance sheet when we think about what do we do next. We have acid rain. We don't need acid ocean."

Despite the sad state of the oceans, Earle is not pessimistic about the future. "There's plenty of reason to be positive. After all, ten percent of the big fish are still there. They're not all gone yet. Half the coral reefs are still in pretty good shape. Despite the fact that we've lost maybe thirty percent and another twenty percent are in bad shape, there's still a chance. I am inspired by individuals." Earle sees promising signs of cooperation around the world.

"Just a few years ago, we had a conference that pulled together representatives from twenty countries. We had seventy different organizations, only a hundred fifty people, that gave rise to something like an action plan for trying to come up with not just 'Woe is me, here are the problems' but [instead] here are some things we can do. And I see conservation organizations working together: World Wildlife Fund, Nature

Conservancy, Environmental Defense, NRDC [Natural Resources Defense Council]—a whole suite of organizations that are really pulling together, or they always have had common objectives but now there is increased motivation to try to get the job done, and that means working with industry because the solutions are there. Or working with governments, working with whatever it takes to find a common ground, to show the connection between a sound environment and a sound economy, a sound environment and health, a sound environment and all the things we hold near and dear, including life itself."

It's not an abstract concept, understanding how we are all in this together, says Earle. "It should be just basic to the way we think. You take care of your home. You take care of the air you breathe, the water you drink, the food that you consume." It ain't rocket science, just logic.

"The thing that worries me is that people seem to be getting increasingly detached from nature. We need to get reengaged, starting with the youngest kids and finding the kid in the oldest of us to realize that we're all connected, first of all, as people, but all people are connected to nature."

How to get reattached to nature? How about simply jumping into the ocean and getting reacquainted? "With knowing, there is the possibility that we can care. If you don't know, you can't care. So getting the word out, getting people to go jump in the ocean and see for themselves what's out there is really critical, but especially kids."

And if a simple splash on the beach is not enough to inspire you, Earle has a more ambitious idea: "I long for the day when you have little rent-a-sub places where you can go get your pickup truck and put a submarine on the back and go off to the dock or the beach of your choice or the boat. It's almost there. You know, there are really dozens of options now that people do have. It's not just for the handful of those who are occasionally lucky enough to penetrate the great depths. The doors are opening. There are passenger subs in Hawaii and Barbados and Bermuda, a few other places, where you can actually buy a ticket and submerge."

CRAIG VENTER GOES FISHING FOR GENES

*We're trying to create a giant database for everybody to use that
gives us a chance to catalog the gene pool on planet Earth.*
—CRAIG VENTER

There's a lot of talk about sending humans to the moon and to
Mars. These planets hold terrific fascination for us, first because
we know so little about them and second because Mars holds the
potential for harboring life, perhaps in a form we have never
seen.

But if you want to explore a place we know virtually nothing
about, you might try looking right here on Earth or more precisely,
right here in our oceans.

"Here we are looking for life on Mars and we know less than one
percent of life on the planet Earth," says genomics pioneer Dr. J. Craig
Venter, the man known more for his work on the human genome
project than for his interest in life beneath the waves. It's not enough
to sequence the genes of humans; Venter will settle for nothing less

than the genes of the entire planet. And most of those genes have yet to be discovered, as tiny microorganisms in the oceans making up 70 percent of our planet. "Ninety-nine percent of the species in the ocean, people don't even know they exist, let alone have a way to characterize them."

Venter has created an unconventional expedition that he says was inspired by the voyages of Darwin's *Beagle* and HMS *Challenger*. And like Darwin, he set sail on the seas in the name of science. Venter and scientists aboard his yacht, the *Sorcerer II*, have been collecting microbes from deep in the ocean, from Maine to the Galapagos, to the South Pacific and back. Their mission: to sequence the genomes of entire microbial ecosystems.

"We're trying to create a giant database for everybody to use that gives us a chance to catalog the gene pool on planet Earth."

His 95-foot yacht circumnavigated the globe, collecting ocean samples every 200 miles. "Our view was why not try and get a first-ever assessment of the diversity in the oceans by doing a complete circumnavigation. We think it's an important part of science education. We're trying to show young people that science can be extremely fun and extremely rewarding in terms of the discoveries that can be made."

Traveling the world filtering seawater sounds idyllic, but it's very fruitful, judging by the results of his first test voyage. Sampling the Sargasso Sea off the coast of Bermuda, Venter and his colleagues discovered a million new genes and thousands of new species of microorganisms. "The biological diversity we found was so vast in the Sargasso Sea; even samples taken a mile apart and a day apart had totally different sets of species in it."

The embarrassment of riches was highly unexpected. "The Sargasso was supposed to be a simple site. It's very low-nutrient, so everybody was speculating there were very few species there. We thought, *We'll start with a simple site and make sure these methods really work well.* What we found: it was far more complex than most ever imagined." The organisms in the Sargasso were breaking the rules of nature as we understand them. These organisms were thriving in abundance without the necessary nutrients—the food—that biologists thought that they would need. What Venter discovered was a system far more complex: the ability of the organisms to capture sunlight and use that energy to overcome the lack of nutrients. "So the thinking is now changing. Perhaps the low-nutrient sites might have more diversity" than sites where there's an abundant source of food for the organisms.

"It means we've not really understood at all the biology of the oceans and that sunlight probably plays a far greater role in the most basic life forms than anybody ever imagined."

It is common for biologists to collect live marine organisms from the ocean and try to keep them alive in the lab long enough to understand what contributions they make to the oceans' ecosystem. But this technique will never work for the tiny microbes in the oceans. They do not readily survive in captivity. Venter hopes to overcome this limitation by using pioneering methods of gene sequencing he developed for the human genome project and applying them to unlock the genetic secrets of these ocean organisms. No need to grow the organisms in the lab; all you need is their genes.

"We take all the DNA from a section of the ocean—in the case of the Sargasso Sea, two million sequences—and we analyze it." The highly successful technique, tested out on the Sargasso Sea organisms, will be applied to the rest of the microbes sampled around the world.

THE OCEANS: A SOLUTION TO GLOBAL WARMING?

In this era of renewed interest in global warming, Venter and his colleagues at the Institute for Biological and Energy Alternatives feel that the ocean creatures may hold a key to understanding and solving the role of the greenhouse gas CO_2. Scientists know that the ocean is one of the most important absorbers of CO_2. It plays an essential role in how nature moves carbon in and out of the atmosphere. "I think one of the biggest findings we had" in studying marine organisms "is all these new photoreceptors," genes that allow plants and organisms to soak up sunlight and inhale the CO_2 to make food and energy. "Maybe half or more of the organisms deal with the sunlight for their source of energy or metabolism versus what was thought to be only a tiny handful of the organisms doing that."

Venter believes that some of these organisms are so exciting that they're going to give us a new view of life. "For example, we're finding that a theme is used over and over and over again in evolution. It's not like we have a million different solutions for each problem. These cells found something that worked really well. Each new cell, each new life form that came along picked up those same elements, modified them a little bit, but didn't change them much over billions of years of evolution. So we think we can actually get down to maybe what was the basic working set in the gene pool that all life derived from."

PART VIII

—

SCIENCE AND RELIGION

CHAPTER TWENTY

FITTING GOD INTO THE EQUATION

I assert that the cosmic religious experience is the strongest and
the noblest driving force behind scientific research.
—ALBERT EINSTEIN

Not long ago, I moderated a panel discussion at the TriBeCa Film
Festival in New York. The festival doesn't normally invite science
journalists to host a panel discussion, but the Alfred P. Sloan Foundation had sponsored a prize for screenwriters who had penned the
best screenplay with a science theme, and I was asked to bridge the
science and the arts divide.

Sitting on the panel, beside the usual film folks, was Dr. James
Watson, the famous codiscoverer, with Dr. Francis Crick, of the
three-dimensional structure of the DNA molecule, 50 years before.
Not only had they gone on to win a Nobel Prize for their discovery
but also a film about the effort was made in Great Britain some years
later starring Jeff Goldblum as the notorious Dr. Watson. (The
science–arts connection on the panel . . .)

The public was invited to sit in the audience and ask questions, and having Watson right in front of them, perhaps the most famous biologist in the world, was an opportunity hard to resist. One questioner in particular was most memorable. She walked up to the microphone, politely introduced herself to the panel, and with great reverence quietly directed a question to Watson about what motivated him in his research. "Did your religious beliefs influence your work?" she inquired very politely. Little was she—or the rest of us prepared—for the tirade that followed.

Watson lashed out at the poor woman for bringing God into the picture. His voice rising to the occasion, he lectured that the main reason he chose to unlock the secret of DNA was to take God *out* of the picture. He would show how life is created without the need to include the deity. How dare she insult him in such a manner? And in a torrent of insult, profanity, and histrionics, Watson proceeded to verbally beat this woman to a pulp. Crushed, she retreated to the safety of the audience, silenced and in a state of shock and awe.

Over the course of the next few months, I would have the opportunity to interview Watson many times and see a similar spectacle: Ask him about religion and he flies off the handle. (After watching this repeatedly, I've come to the conclusion that Watson is a showman and this is his act.)

Other scientists, such as Steven Weinberg and Richard Dawkins, are as opinionated about science and religion as Watson. And they take no prisoners either, though they lack the flair for the dramatic that Watson has cultivated. All of which raises the questions: Just how much does a person's faith influence his or her work in science? Can a person with serious religious beliefs also be a serious, respected scientist?

These questions are especially relevant in the current religious climate. Has there been a time, in the past hundred years, where science and religion have clashed more forcefully than we are witnessing now? Not since the Scopes trial in 1925 have we seen such vocal clashes between opposing forces of science and religion.

From President George W. Bush and his "the verdict is still out" when it comes to the validity of evolution, to the Pennsylvania judge who wrote the definitive ruling against teaching creation in biology class, to the debate over embryonic stem cell research, science and religion have been in the news almost every week. Which again highlights the issues: Can science and religion coexist? Can religious scientists also be good scientists? What does religion inform scientists about their work? What does research inform scientists about their beliefs?

The answers are not simple. Over the many years that I have interviewed scientists, I have never seen a wider range of views by scientists and theologians about religion and science than I do now. The opinions tend to fall into three broad categories:

• Religion is antiscience, and as long as the two maintain a distance, the better. Many who hold this view are atheists.
• Extreme fundamentalist views of religious beliefs, where the bibles, of many faiths, are taken to be literally true and thus cannot accommodate modern scientific beliefs such as evolution, age of the Earth, and the equality of the sexes.
• Religious views are not in opposition to science and can bend as science makes new discoveries. An interesting corollary to this places God outside of the natural world so that God and science don't intersect except in unusual circumstances called miracles.

I've collected some of the more cogent and vocal lines of reasoning to illustrate the three views above and to give you a flavor of the various and creative ways that scientists have come to terms with religion in their lives, both welcoming its influence and forcefully shunning it.

Just as religions provide you a buffet of choices out of which to find a belief that fits your worldview, so do the following personalities, opinions, and ideas, out of which I'm sure you'll find some with which you agree. Maybe even a few!

The debate about religion and science is sure to be a topic for years to come, as it has been for the last four centuries.

THE FLAMETHROWERS

On one end of the spectrum lie those scientists who show little accommodation for religious points of view. James Watson, mentioned above, is a good example. But he is not alone. "I grew up without religion, and as I've gotten older, I've become more and more inclined to dislike the influence of religion in the world and to hope that the human species will eventually outgrow it," says Dr. Steven Weinberg, professor in the Department of Physics, University of Texas at Austin, and winner of the 1979 Nobel Prize in physics. Weinberg is an avowed atheist. "I'm in physics because I love physics. But I do hope that the advance of science in general and physics in particular will in the long run continue, as it has in the past, to weaken the hold of religion on people's minds."

When I first heard Weinberg's acid critique of religion and those who are believers I was shocked. It is not unexpected that some scientists harbor these beliefs. But here was someone who, like Watson, hardly sugarcoated his message. And he never shied away from telling just how he felt about it. For example, while Weinberg said in a 2005 appearance on *Science Friday* that he is in favor of a dialogue between science and religion, he said that such an exchange would not be a constructive dialogue. "There is no conflict between science and religion on the level where religion is simply a matter of ethnic identification or habitual practice or aesthetic enjoyment. But when it comes to belief in things being true or not true, as, for example, that there is something beyond the physical universe, that there is an afterlife, that there is a deity, it's on the level of belief that science and religion come into conflict. That's not all of religion, by any means, but it is an important part."

The most famous case of religion intruding on science and seeking to influence its findings is the one involving Galileo reporting that he had seen the moons of Jupiter revolve around that planet. That put him at odds with the Catholic Church and he was forced to retract what his eyes told him to be true on the threat of excommunication by the pope. Weinberg says that as science has advanced through the centuries, it has caused religion to retreat from areas in which it used to make very definite statements of belief about science and nature. "I think that's a good thing." But, he says, it still has not been a full retreat.

"There are still idiots out there in the world who think the world was created six thousand years ago and that all the species were created at the same time." He is willing to admit that those folks represent a minority opinion, though.

THE MODERATES

Dr. John Haught doesn't believe that the purpose of religion is to compete with science. "If religion were in the business of trying to give out scientific information, then surely it's going to compete with science, and there could only be a retreat if that was the function of religion," says Haught, professor of theology and director of the Georgetown Center for the Study of Science & Religion at Georgetown University in Washington, D.C. "I grew up as a Roman Catholic and I also came from a family of scientists, and so I've always had a love of both science and theology, and as a Catholic, I've never had any problem reconciling the two."

Here is the typical middle ground of the moderate view: "In many ways, the coming of science has delivered religion from the moonlighting job of trying to provide anything similar to scientific information. Religion is in the business of doing something quite, quite different. And as such, there can be no conflict. In a very broad sense, you could say religion is belief in something of ultimate importance, and even the scientist has religion in that sense.

Or in a more narrow sense, you could say religion is belief that there's some sort of incomprehensible mystery that surrounds the universe. Einstein himself accepted religion in that sense of the term."

Einstein's God *was* a spiritual God, one whose secrets were there to be discovered. A "cosmic religion," as he called it. He dismissed the idea of a "God conceived in man's image, so there can be no church whose central teachings are based on it."

"Where I think the rubber hits the road and things really get sticky," says Haught, "and I think Steven Weinberg would probably agree with this, is where religion is understood as belief in what Steven calls an interested God or a personal God. I think what religion is attempting to do in using the symbol of a personal God is to get across the belief or the conviction that the universe ultimately is undergirded by a principle of care, by a principle of intelligence, by a principle of meaning. And you know, science doesn't do anything like that. Science is in the business of gathering a much more limited kind of information, and as such, I don't think it can conflict with religion as I've understood it."

Does Weinberg agree? "There still is perhaps not a conflict but a tension, as Susan Haack, the philosopher, calls it between science and religion, on two levels. It's certainly true that science is never going to disprove the existence of an interested God or an afterlife. But when you learn the way the world works, as you learn more and more about the laws of nature and their chilling, impersonal quality, as you learn more and more about the irrelevance of human life to the general mechanism of the universe, the idea of an interested God, of a path that has been laid out for human beings, of a cosmic drama in which human beings are playing the starring role, becomes increasingly implausible; not disproved but just implausible.

"But there's even a deeper level on which there's a conflict or a tension between science and religion, and that is that science demonstrates a mode of knowing, which, as has been said, is culture-free, does not look back to sacred texts. No one today, in trying to settle an argument about general relativity, would quote Einstein as an authority, because we progress. He's a hero of ours, but he's not a prophet. He's not someone whose sacred writings we have to take seriously.

"This is rather different from the world's religions. Even on the elevated, and I would say rather exceptionally enlightened level, the tension is there, and I think it will continue to weaken the influence of religion; even though certainly, I would never argue that science is all we need. Yes, there's a lot more to life than science. There's a lot of things science can never give us. But I just hope that the people will grow up and stop looking for the other things that science can't give them in the supernatural."

DOES SCIENCE MAKE IT HARDER TO KEEP RELIGIOUS FAITH?

Ever since the enlightenment, ever since astronomers started pointing telescopes at those moons of Jupiter or assaying the age of the rocks and seeing them to be quite older than the Bible says, scientists have been wrestling with the problem of how to reconcile what new discoveries tell them to be true with religious views that may be in conflict. Can religious views be "updated" or "modernized" to conform to constantly changing views of the world?

"Judaism has gone through many different changes and understandings of science," says Laurie Zoloth, director of bioethics for the Center for Genetic Medicine and professor of medical ethics and humanities and religion at Northwestern University's Feinberg School of Medicine in Evanston, Illinois. "It's been a witness to extraordinary changes in our understanding of the molecule, of the self, of the cell, of the universe. And in fact, the religion is strengthened by a more acquisitive and inquiring pursuit of what the real world looks like. It's never diminished the understanding of what it is to be a Jew or the duties that a religious Jew carries forward into the world."

So perhaps some religions, or rather some liberal or more flexible sects of those religions, find it easier to accommodate changing scientific views of the world. But not all. "The general question about whether science makes it harder to keep religious faith is the wrong question," says Susan Jacoby, director of the Center for Inquiry—Metro New York and author of *Freethinkers: A History of American Secularism*. "The question is: What kind of religious faith are you talking about? You cannot, for example, believe in the scientific method or accept the theory of evolution and also believe that the universe was created in seven days, as fundamentalist Christians do. You can certainly believe in the theory of evolution and believe that God set the universe in motion. What you cannot do is believe in the scientific method and believe in a fundamentalist, literalist form of religion, whether it's fundamentalist Judaism, fundamentalist Protestantism, fundamentalist Catholicism, or fundamentalist Islam."

Zoloth disagrees. She sees it possible for religious dogma to bend to new interpretations that do not conflict with scientific understanding. "Let's just look at the example that you just gave, Susan. You can be a Jew who believes in traditional texts and understand that the world could be created in seven things that are translated as days, but aren't Monday, Tuesday, Wednesday, Thursday, Friday, and Saturday. Right? So then there's a linguistic and normative account of reunderstanding what we mean by a day. And that's the move that's made in traditional texts and even in traditional communities." In other words, a day can be much longer than 24 hours; perhaps it can be eons, if you want to reconcile it with science. Zoloth says a close reading of Genesis notes that the concept of "days" being periods of 24 hours doesn't exist early on in the creation story. There are "things called days before there's suns and moons to define them, so there's got to be some other way of understanding it. And oftentimes people have said, 'Look, there's two different ways of thinking about the word, the text and that's been done since the very beginning of textual interpretation."

FUNDAMENTALISM

What about rigid religious beliefs that don't bend?

"Everybody is talking very nicely, as though there doesn't have to be a conflict because you're talking about people who want to make accommodations, who want to make the days a metaphor for something else," says Jacoby. "But I don't see how any of us can deny, whatever faith we have: there's a very large percentage of people in this country and this world who do not want to make those accommodations, who are not interested in those accommodations, and who in fact say, 'No. If the Bible or my interpretation of the Koran doesn't say this literally, it cannot be true.' And this is why we're having so much controversy. I think we face a situation, a dangerous situation, where a lot of people refuse to allow their religion to be modified by secular knowledge at all."

Sounds very much like Weinberg's view. "I think what we're hearing is a very enlightened view of religion. It's one which I would argue does represent a retreat in the sense that if these opinions had been expressed about the fact that religion does not compete with science in the description of the world around us, the physical world, these are people who would have been burned in the Middle Ages and might today have quite a lot of trouble in many Islamic countries."

Islam is a very interesting example. Islam had a "golden age" of science and math. Muslims were encouraged by the Prophet Muhammad to seek new knowledge, to search for it even as far as "China." As for diseases, scientists were encouraged to find cures for which Allah had provided. While Islam has a long tradition of welcoming and creating advances in science and math, Dr. Ebrahim Moosa, associate research professor in the Department of Religion at Duke University and codirector of the Center for the Study of Muslim Networks, says today, "The political conditions in different parts of the Muslim world create certain levels of anxieties that, in my view, block that kind of flourishing of intellectual thought, and you have forms of political repression as well as a very atrophied version of religion that is very loud, and that also creates further repression of critical and scientific thought."

GOD VERSUS NATURE

One way that many scientists have come to grips with their beliefs is to put them in separate boxes. We don't need to modernize our religious views, no need to jump through hoops to make text match science because there is no conflict: God and nature coexist but don't interact.

"If God has any meaning, God is outside of nature, not bounded by nature," says Dr. Francis Collins, director of the National Human Genome Research Institute at the National Institutes of Health in Bethesda, Maryland. You may remember that he led the public effort to decode the human genome.

At first glance, this might seem to be a contradiction. Isn't God part of nature? Doesn't the Bible tell us that God created "the heavens and the Earth"? Or is God supernatural, outside the boundary, as Collins posits? Think about it. When you look around at nature and believe that God has created it, isn't it natural to assume that God is part of nature? But what if you can put the two worlds in different "boxes"? Because if you can put the deity outside of nature, into a box of its own, then you don't have to have a conflict between religion and nature. If God is supernatural, the logic goes, and science can explain only the natural world, then you don't have a conflict. You can believe both in science and in God.

"So while science is the only really reliable way to understand nature, science is relatively powerless to help us with the question of whether God exists, and if so, what he is like," concludes Collins. And others agree.

"The doctrine of creation is trying to get across the point," says Haught, "that the universe is grounded in an ultimate reality, that the universe is a gift from something other than nature itself, that it's trying to answer the question 'Why is there anything at all rather than nothing?' rather than trying to give us information, scientific information, about how our universe came into being."

"I believe there are many, many things in human life which transcend the explanatory levels of life, and religion is precisely dealing with those other aspects which have little to do with explaining natural phenomena," says Dr. Varadaraja V. Raman, professor emeritus of physics and humanities at the Rochester Institute of Technology in Rochester, New York. "It seems to me that the two kinds of truths—religious and scientific—belong to entirely two different categories, ontologically speaking. One, religious truths are grounded, anchored, to geographical, historical, and cultural truths, which are enormously important, whereas scientific truths transcend these and are, by definition, space and time invariant and culture invariant. And that is why efforts to bring the two together often fail.

"I come from the Hindu tradition, where religion means something very different, because it is not based on our beliefs but on our experiences, the idea being that there is something beyond the physical world, which we experience in different ways. So I have always been introduced into this mystical dimension, as it were, or the spiritual dimension, more exactly, and my love of physics has been there since my high school days, and so I'm as much a physicist as one devoted to the other dimension of human life, mainly the spiritual."

Dr. Owen Gingerich is the author of *God's Universe*. He is senior astronomer emeritus at the Smithsonian Astrophysical Observatory and research professor emeritus of astronomy and the history of science at Harvard University. He is one of the most respected historians of science in the world. He does not believe that by studying science one can find proof of the existence of God. Rather, for him, "the universe makes more sense with this understanding." And he too has no trouble separating science from religion: "A creator-God gives a coherent understanding of why the universe seems so congenially designed for the existence of intelligent, self-reflective life. You don't even ask whether God exists because God is not an object, and therefore it's the wrong question."

"And so while science is the only really reliable way to understand nature, science is relatively powerless to help us with the question of whether God exists, and if so, what he is like," says Collins. "Is it not immediately apparent that those questions [about God] are not scientific questions. And if you seek answers to those, you have to find them in another place. And that's what, in very much a strong way, drove me toward considerations of spirituality after having been living for quite some time in a very materialist perspective as a graduate student."

Collins's conversion occurred as a medical student "when after watching people lean on their faith for great support at times of considerable challenge—and many of them facing certain death—I got curious about what this was all about and how people could derive such strength from it. And seeking to shore up my atheism by having a better defense against that kind of faith, I accidentally converted myself."

So does Collins find that his science life impacts his religious views?

"Oh, yes, it does. Stephen Jay Gould, of course, had this famous argument: that the spiritual and the scientific world views are both valid, but they ought to avoid appearing in the same room or the same brain at any given instance because an explosion might occur—the nonoverlapping magisterial model.

"That doesn't work for me. For me, being a scientist who's also a believer is a wonderful, comforting, harmonious experience. So that as a scientific discovery looms into view—and we scientists have the chance to do that from time to time—it is both a remarkable moment of realizing that you've discovered something that no human knew before, but God knew it. And so you are experiencing discovery and also a chance to glimpse just a little bit of God's mind. And for me that is just a privilege and a wonderful experience not to be missed."

Sentiments like that set Gingerich to quoting Scripture. "I think the most important verse in Genesis 1 is that 'God created man in his own image. Male and female created he them.' And what would that mean? I would say it's creativity, it's conscience, it's consciousness. By being a scientific researcher and discoverer, we are, in a kind of sense, linking ourselves with God because the rationality of the universe, the fact that we can find these regularities and so on, is part of this logical structure that is God."

Collins points to our old friend Galileo "who obviously got into some hot water here but who was also a strong believer as well as a wonderful scientist—wrote this wonderful sentence: 'I do not feel obliged to believe that the same God who has endowed us with sense, reason, and intellect has intended us to forego their use.'

"So I agree with what Owen is saying. One of our greatest gifts in the image of God is an intellect and the ability to explore the natural world, God's creation. And I think God expects us to use that gift and is not threatened by what we discover."

How do these beliefs square with the rise of intelligent design, the effort to teach the biblical story of creation in biology class, as a challenge to evolution? Collins believes those intentions are misguided.

"As a believer and a scientist, I am quite troubled by particular interpretations of Genesis 1 or of the failings or positive revelations of evolution that pit believers against scientists in a way that, I think, is really unnecessary. Intelligent design as it is currently being proposed is a very special kind of view about evolution—namely, that evolution was not sufficient to account for some of the really complex molecular machines that we find inside ourselves, like the bacterial flagellum, for instance, which is a favorite example.

"Proponents of [intelligent design] argue that only the intervention of a supernatural force would make it possible for such machines to come into being, because they are constituted of so many subunits. And if you knock any one of them out, you lose the function, that evolution would never be able to put this all together, since you wouldn't gain any advantage until the very last step.

"That really flies in the face of what we're learning about how such machines are built up bit by bit from smaller subunits that had other functions. And so I fear intelligent design is a God of a 'gaps theory,' which puts God in a box and makes, ultimately over the course of time, a theory that is likely to collapse before too many more years go by, and in the process does no damage to science but actually may do damage to faith."

Collins believes that the tactics used by creationists may backfire "because young people looking at the evidence may soon conclude that if you're going to be a believer you have to basically deny scientific facts that are well substantiated. And that's a terrible choice to ask somebody to make and a totally unnecessary one, although I don't question the sincerity of those who are promoting the intelligent design perspective or those who believe in young Earth creationism. These are people who have really seriously attached themselves to something that they feel defends them against what they perceive as a very atheistic perspective that seems to be coming out of some quarters in science, a perspective which, I might add, is unnecessary and also destructive."

"I like an example given by John Polkinghorne, who's an eminent physicist and also an Anglican priest," adds Gingerich. "He asks the question: Why is the water in the teakettle boiling? Well, you can explain it by the heating going in the bottom, the molecules rushing faster and faster, some of them finally flying out the top. The water begins to evaporate. But you can also answer the question: Why is the water in the teakettle boiling? Because I want some tea.

"These are examples of what Aristotle would call efficient causes or final causes. Science works by efficient causes. Explaining how things work, that's where it's made its great success, and that is the kind of answer that can get your spacecraft to Jupiter, for example. Explaining it in terms of the molecules or explaining evolution in terms of DNA, genetics, mutations, selection, that's an efficient cause. A final cause may be that this is part of a divine plan for how the cosmos is supposed to turn out.

"But I don't think you can swap one of these for the other. You can't replace the teaching of evolution in biology classes with intelligent design, even though at some level both may be true."

A MIRACLE: GOD INVADES THE NATURAL WORLD

The conversation gets even more interesting with the subject of why life chose Earth as its home planet. Why have we not so far found life in other parts of the universe? Could it be that the conditions for life on Earth are so precise as to make it a unique place in the universe? And if so, is it a sign of the hand of God? Gingerich does not rule out the possibility.

"Even Stephen Hawking in his book A *Brief History of Time*, points out that this precise tuning of these physical constants, sometimes referred to as the 'authropic principle,' seems to have major theological implications. And I think that's actually a fairly interesting argument that most people aren't aware of. Why should it be that the gravitational constant has exactly the value that it needs to in order for, after the big bang, actual coalescence of stars and galaxies and planets to have occurred instead of having things continue to drift off infinitely or else come back together quickly in a big crunch? And I think actually, considering the options, it's a pretty good option" that God wanted it that way.

Let's follow this line of reasoning a bit further. Does that mean that a scientist who believes in the Bible, where lots of miracles happen, would also tend to believe in miracles in nature?

"The big question," says Gingerich, "is whether you're willing to accept the reality of God as a supernatural being. That is the decision that all of us at some point are faced with having to make, although many of us try to avoid it as long as we can because maybe the whole question of faith makes us uncomfortable. But once you've come to that point, as I did at age twenty-seven, of accepting the possibility—in fact, the reality—of God who is outside of nature, then that solves the miracle problem for you fairly quickly. Because if God is real, there's no reason that God could not occasionally stage an invasion of the natural world, which would to us appear as a miracle.

"Now, don't get me wrong. I've never seen one. I don't think they occur on any sort of regular basis. They may, as C. S. Lewis describes, occur at those great ganglions of history where it suits God to make a point of some sort. They're not randomly shaken into human experience.

"But I don't think one needs to say, even as a rigorous show-me-the-data scientist, which I am, that miracles are impossible once you've accepted the idea that God exists."

UNWRITTEN TABOO?

How do other scientists see the religious views of scientists like Collins and Gingerich? Collins says they are generally well accepted.

"Robert Jastrow, the noted astronomer, began his own book, *God and the Astronomers*, with this rather famous sentence: 'When a scientist writes about God, his colleagues assume he's either over the hill or going bonkers.' And I hope I'm neither of those. But I take comfort in the fact that surveys would tell you that forty percent of working scientists believe in a personal God to whom one may pray in expectation of an answer.

"They're fairly quiet about it. I think there is sort of an unwritten taboo that you don't discuss this in scientific circles, but there's a lot of us who share that same view, including, I might add, some prominent evolutionists over the course of the last century, like Asa Gray, who was the major defender of Darwin in the U.S., and Theodosius Dobzhansky, probably the greatest intellect of the early part of the twentieth century in the area of evolution and who found no controversy or conflict at all between being believers and also those who clearly saw in evolution an answer to a lot of questions about how living organisms are related to each other."

SPEAKING OUT

In writing books about their faith, in being the cover stories of national magazines, as Collins has, both he and Gingerich have chosen to "go public" with views that they had more or less kept quietly to themselves, or at least had not discussed under the glare of the public spotlight before.

"It's a bit like taking a public bath, I suppose, but I think Owen would probably resonate with this as well. When you look around us today, particularly in the United States, there seems to be such a battle going on, with the extreme positions dominating the stage. Some of those extreme voices come from our own colleagues who pronounce in shrill tones that evolution and other discoveries of science have rendered God no longer necessary and that any thinking person should now become an atheist. On the other pole, one hears pronouncements from fundamentalist perspectives that science cannot be trusted if it disagrees with an ultraliteral interpretation of Genesis, and anybody who disagrees with them is not a true believer.

"These are troubling times, and if there are those of us who have not arrived at either of those extremes but actually inhabit the middle ground in a very comfortable way, for whom the spiritual and the scientific world view are not only compatible but complementary and comforting, should we not be speaking up about that?

"Because a society in the future that abandons science or abandons faith is not a society that's necessarily a healthy one. And if we have the chance to try to preserve both of those in a healthy way, then I think we have some obligation to do so. So yeah, there are risks in being outspoken about this, and certainly Owen and I have encountered some negative reactions about it—although not much. I think most scientists are respectful, if perhaps in some instances a bit puzzled."

Gingerich agrees. "We're looking for a kind of middle road between two fundamental extremes. You can have fundamentalist scientists who are so absolutely sure they understand it all and who are hard-core atheists, and you can have fundamentalists on the religious side who are prepared to take a literalist reading of the Scriptures that has not been borne out historically. I have been speaking about these issues for some time, but I think there needs to be a kind of a middle voice in this, and I've tried to represent that field."

As for atheists, Collins believes they must have even more faith than many people who believe in God. "They have to have faith in their own intellect's ability to know so much that they can exclude the possibility of God categorically, which seems to me the greatest statement of faith, or perhaps hubris and arrogance, that one could imagine. So yes, faith, but in what? As I look about myself and the culture we live in and the world we live in, a world without the kind of noble intentions that arise, many times, out of people's hearts in the consequence of their faith, a world that misses out on a Mother Teresa or an Oskar Schindler, a world where science has to go on in a completely materialist way, does not sound like the kind of world of wonderful humanity and nobility of humankind that I hope will be evolving over the many decades to come."

"I find it really very offensive to be told that if there is no God, then we can have no moral principles," says a perturbed Weinberg. "I don't believe in God, and I think I live quite a moral life. I think in fact our view of what moral principles are taught by religion has been evolving through the change in morality in society.

I mean it used to be thought that there was no conflict between Christianity and the existence of slavery. Today, many people think there is.

"I think the evolution has not been an evolution of religion but an evolution of the general morality of society. And I certainly don't like to be told that because I don't believe in God that I therefore must be an immoral person."

EVOLUTION: STILL UNDER ATTACK

Be it enacted by the General Assembly of the State of Tennessee, That it shall be unlawful for any teacher in any of the Universities, Normals and all other public schools of the State which are supported in whole or in part by the public school funds of the State, to teach any theory that denies the story of the Divine Creation of man as taught in the Bible, and to teach instead that man has descended from a lower order of animals.

— STATE OF TENNESSEE, March, 1925

All eight members up for reelection to the [Dover] Pennsylvania school board that had been sued for introducing the teaching of intelligent design as an alternative to evolution in biology class were swept out of office yesterday by a slate of challengers who campaigned against the intelligent design policy.

— NEW YORK TIMES, November 9, 2006

If you thought that the famous Scopes Monkey Trial of 1925, in which a Tennessee schoolteacher, John Scopes, was tried, convicted, and fined for teaching the theory of evolution was the final word in the war between scientists, educators, and creationists over the teaching of Darwin's views in the classroom, you haven't been paying attention.

EVOLUTION: STILL UNDER ATTACK

The acrimonious debate received renewed vigor after George W. Bush declared that when it comes to evolution, "the verdict is still out." His conservative supporters redoubled their attack on science that had been bubbling beneath the surface of local school politics in the 80 years since the Scopes trial. In 1999, the Kansas Board of Education voted to remove evolution from the list of subjects to be tested on state achievement exams. Dissatisfied, Kansas voters in 2000 voted out the school board that had sanctioned that proposal. But the tug of war continued. In 2004, conservatives once again gained control of the school board and voted to include the teaching of "intelligent design," a repackaged version of creationism, in public school biology classes. The pendulum swung once again, in 2006, when pro-evolutionists took control again, in the political tide that returned control of Congress and many state and local elections to centrist Democrats and social ideas.

But it was Pennsylvania, not Kansas, that became the magnet of public attention in the early years of the twenty-first century. In 2004, the school board of rural Dover, Pennsylvania, mandated that teachers include the reading of a statement about intelligent design before teaching evolution. It became the first school in the United States to make such a requirement and in doing so, brought the full attention and focus of scientists, educators, and the media to this small community.

In revolt, some teachers resigned rather than condone the attack on evolution. Incensed, eleven parents challenged the decision in court. A dramatic trial in the state capital, Harrisburg, brought the issue to a surprising conclusion when U.S. District Judge John E. Jones III issued a 139-page ruling saying, among other things, that intelligent design is not science but religion and had no place in the classroom. In addition, Judge Jones, in a broad and scathing decision, said that the requirement violated the First Amendment to the Constitution and "cannot uncouple itself from its creationist, and thus religious, antecedents." Judge Jones was appointed to the U.S. District Court bench in 2002 by President George W. Bush.

What follows are some of the discussions that framed the issues, focusing specifically on the Dover case. What you will not find here is a debate about the validity of intelligent design. It is a conscious choice, on my part, not to have one. Proponents of intelligent design intend to wedge their theory into school curricula by creating an artificial and clever construct: that scientists are unsure of the truth of evolution themselves, that there is a real "controversy" among scientists about its validity. That teachers would be doing a disservice to their students by not "teaching the controversy."

But the fact is that there is no controversy to teach. Sure, scientists disagree and debate just how evolution is occurring, but as a rule, they do not disagree about the basics. Therefore, for me as a journalist to conduct a debate about the issue would be, in effect, to "teach the controversy" and grant the creationists a victory.

What I choose to do instead is to listen to the voices of people making the decisions about teaching intelligent design and evolution: politicians, school board members, science teachers, and the rulings of courts deciding the issues who all have valuable personal and professional reasons to be heard. This is an issue that is not going away anytime soon.

THE DOVER SCHOOL BOARD CASE

On the issue of evolution, the verdict is still out on how God created the Earth.

—GEORGE W. BUSH

Of all the recent challenges to evolution, none is more important—or more surprising—than the Dover School Board case.

In October 2004, the Dover School Board near York, Pennsylvania, voted 6 to 3 to require science teachers in its public schools to include in their biology classes an idea called intelligent design. Proponents of intelligent design say that some parts of life are just too complex to be the products of evolution. An intelligent designer probably had to become involved. Evolution alone, they say, doesn't explain the origin of species. The intelligent designer, though never specified, is not thought to be a space alien but rather a supernatural being, such as God.

A note on the school board's Web site noted that the biology curriculum was updated to include the following statement: "Students will be made aware of gaps, problems, in Darwin's theory, and of other theories of evolution, including but not limited to intelligent design. The origins of life is not taught."

Another section said, "Because Darwin's theory is a theory, it is still being tested as new evidence is discovered. A theory is not a fact. Gaps in the theory exist for which there is no evidence. A theory is defined as 'a well-tested explanation that unifies a broad range of observations.' Intelligent design is an explanation of the origin of life that differs from Darwin's view. The reference book *Of Pandas and People* is available for students to see if they would like to explore this view in an effort to gain an understanding of what intelligent design actually involves. As is true with any theory, students are encouraged to keep an open mind. The school leaves the discussion of the origin of life up to individual students and their families."

A majority of the school board in Dover, Pennsylvania, agreed, and voted that students there should be taught in science class ideas contrary to what the overwhelming majority of scientists believe about the origin of species.

Jeff and Carol Brown were members of the school board who could not support the decision and chose to resign. "After ten years on the board," explained Carol Brown, "I realized that I was at a point where I could no longer represent the interests of all of the members of the community, because that's what I was elected to do. Not the interests of a small group or an individual but the interests of all of the members of the community. And I was no longer able to do that."

The controversy started over the selection of a new biology textbook, which was opposed by the person in charge of the curriculum. "He cited sixteen mentions of Darwin and Darwin's theories in a 1,400-page textbook," said Carol Brown, "and he said that he would not present it to the board for approval, and he thought that we should look further." Brown was surprised at the decision, because "at no point in the textbook, in no chapter of the textbook, is there any mention of the so-called controversial theory of the descent of man, which Darwin promulgated. The discussion centers about Darwin's theory of natural selection, which is, in essence, survival of the fittest. If a species cannot adapt, then that species dies out. And that's all we have ever taught in Dover, and that is what is mandated by the state curriculum."

Yet despite years of teaching Darwin, the school board was now deciding to eliminate the standard biology textbook and substitute it with a book called *Of Pandas and People*, which espouses the idea of intelligent design. "I see this as an attempt to put a specific religious viewpoint into the classroom that is contrary to the 1987 Supreme Court ruling."

When Ms. Brown objected to the book, the school board frequently called her faith into question. "I would never presume to ask anyone what his or her faith is. That's of no importance in serving on the school board. We're there to represent all of the members of the community and all of our students, and that includes students of all faiths. And I was asked, actually, on three separate occasions whether or not I was a born-again Christian, whether I had been 'saved.' And although I didn't go off in anger, I was very offended by that, because that is a matter between me and my God, if you will, and my pastor."

Jeff Brown objected to the inclusion of intelligent design on scientific grounds, noting that the board's own statement read, "A theory must be well tested." He noted, "Intelligent design is not well tested. It hasn't been tested at all. And the fact of the matter is, it cannot be tested. I'm waiting for someone to submit as an experiment, any kind of experiment, that would prove or disprove this intelligent intervention. It's a supposition that's essentially unprovable. They've just sort of simply thrown it out there and said, 'Well, this is a countertheory.'" That notion did not sit well with Brown. "Yeah. And the fact that the universe came from a giant Cracker Jack box is another possible theory, but you can't prove that one either. It's not science."

Jeff Brown's views were not received very well by the school board. "Basically, I was ostracized, and people that I considered to be friends stopped talking to me." The community, the people of Dover, treated Brown more sympathetically. Overwhelmingly Christian, Dover residents nevertheless were not all of one mind when it came to teaching creation in the classroom. "I know an awful lot of Christians who don't have a problem with evolution, but this is a segment of the Christian community that does, and their feeling is that by teaching evolution and not teaching their side—and that's their word, their 'side'—we're not giving the kids a well-rounded education. My understanding of the law is if we're going to teach their side, we have to teach everybody's side."

Carol Brown did offer that solution, repeatedly over the years. She had no problem bringing religion into school classrooms but not into biology class. "Why not offer a comparative religions course in the high school, to give students an opportunity to learn about the religions of the world, the beliefs of each one? And as I pointed out, the underlying factor in every single faith is basically what we call the Golden Rule: Treat other people the way you want to be treated. And that was not well received."

Shortly after the Brown resignations, the parents of Dover schoolchildren took action. Eleven of them took the school board to court, in the first challenge against a public school district in federal court to the required teaching of intelligent design. The parents were backed by the American Civil Liberties Union and Americans United for Separation of Church and State. Supporting the school district was the publisher of the textbook *Of Pandas and People*, the Foundation for Thought and Ethics. The trial, which began on September 25, 2005, lasted 21 days with Judge Jones taking a month to issue a ruling. Before Judge Jones's ruling on December 20 came the first surprise: All eight school board members who voted to include intelligent design were voted out of office in a November 8 school board election.

"That was quite something, wasn't it?" Nick Matzke of the National Center for Science Education watched all six weeks of the trial as a consultant for the plaintiffs' legal team, giving advice on the science. "I was just sitting there observing, and so I did not have an idea of how things were going out on the ground in Dover, in the actual community. And so when the election result came out, I was pretty amazed." The overwhelming vote against intelligent design, says Matzke, represents what happens once people pay close attention to the issues. "Whether they want to or not, once they get really familiar with intelligent design, people realize what's really behind it. They realize it's really just creationism in disguise."

Obviously Judge Jones wasn't fooled either. On December 20, 2005, Judge Jones issued a 139-page decision. It was a scathing attack on the idea of teaching intelligent design (ID) in biology classrooms. Speaking for the court, Judge Jones wrote:

> . . . we have addressed the seminal question of whether ID is science. We have concluded that it is not, and moreover that ID cannot uncouple itself from its creationist, and thus religious, antecedents.

In other words, ID is creationism in another wrapping. Judge Jones went on to attack the notion that evolution and religion are incompatible:

> Both Defendants and many of the leading proponents of ID make a bedrock assumption which is utterly false. Their presupposition is that evolutionary theory is antithetical to a belief in the existence of a supreme being and to religion in general. Repeatedly in this trial, Plaintiffs' scientific experts testified that the theory of evolution represents good science, is overwhelmingly accepted by the scientific community, and that it in no way conflicts with, nor does it deny, the existence of a divine creator.

To be sure, Darwin's theory of evolution is imperfect. However, the fact that a scientific theory cannot yet render an explanation on every point should not be used as a pretext to thrust an untestable alternative hypothesis grounded in religion into the science classroom or to misrepresent well-established scientific propositions.

The citizens of the Dover area were poorly served by the members of the Board who voted for the ID Policy. It is ironic that several of these individuals, who so staunchly and proudly touted their religious convictions in public, would time and again lie to cover their tracks and disguise the real purpose behind the ID Policy.

Very stern words about fundamentalist religion written by a judge appointed by a fundamentalist President Bush. But he had more to say. Not only were several school board members dishonest about their convictions but also the board violated the First Amendment of the Constitution of the United States and the Constitution of the Commonwealth of Pennsylvania.

Our conclusion today is that it is unconstitutional to teach ID as an alternative to evolution in a public school science classroom. Those who disagree with our holding will likely mark it as the product of an activist judge. If so, they will have erred as this is manifestly not an activist Court.

Judge Jones had more:

The breathtaking inanity of the Board's decision is evident when considered against the factual backdrop which has now been fully revealed through this trial. The students, parents, and teachers of the Dover Area School District deserved better than to be dragged into this legal maelstrom, with its resulting utter waste of monetary and personal resources.

Wow. Summing up, the Judge entered an order "permanently enjoining Defendants from maintaining the ID Policy in any school within the Dover Area School District, from requiring teachers to denigrate or disparage the scientific theory of evolution, and from requiring teachers to refer to a religious, alternative theory known as ID."

Barbara Forrest is an expert witness in the trial whose testimony helped persuade the judge that ID is creationism relabeled. Judge Jones specifically remarked that Dr. Forrest's testimony and exhibits "provides a wealth of information, a wealth of statements by ID leaders that reveal ID's religious, philosophical and culture content." The defense tried and failed to have her barred from testifying at the trial. Dr. Forrest is a professor in the Department of History and Political Science at Southeastern Louisiana University and coauthor of *Creationism's Trojan Horse*. Dr. Forrest is "not totally surprised" by the ruling because, she says, that during the trial "whenever he would issue rulings in response to motions, they were very thoughtful, very carefully done. But I think it really sets a benchmark for judicial excellence and integrity, especially with respect to this issue."

But what will be the impact of this ruling on other schools where teachers and parents are fighting to keep intelligent design out of science classes? Forrest says Judge Jones's ruling will send a strong message to some school boards, perhaps most, but not to others.

"But one of the things that we know from the history of creationism and the religious right in general is that they tend not to pay attention to court rulings. We thought that in 1987 with the *Edwards vs. Aguilar* ruling that came right out of my own state of Louisiana that that would put an end to the problem of creationism in this country, and obviously it did not. The good thing about Judge Jones's ruling, though, is that it didn't leave the intelligent design/ creationists much room to morph. What creationists usually do, in response to their losses in court, is that they change themselves into something a little bit different, but I don't think they have much room to do that after Judge Jones's ruling.

"You also have to recognize that the creationism issue is not based on evidence. It isn't based on reason. It's based on an uncritical acceptance of certain religious doctrines that are not very thoughtfully held. And so when you get a position that is not based on evidence and rational appeals, you get people who are going to ignore court rulings because they are motivated by religious zeal."

People who have made up their mind tend to pick and choose facts from an argument that back up what they believe. If you believe the Earth is flat, no amount of satellite imagery taken from space will convince you. If you believe that the Earth is only 6,000 to 10,000 years old, then no radioactive dating mechanism is going to change your mind, either.

Dr. Forrest agrees. "One of the amazing things that Judge Jones points out in his ruling—he talked about this in detail—is that when the Dover School Board adopted this policy endorsing intelligent design, they themselves had no understanding of it. That was amazing. Not only do they not know the science—the supporters on the Dover board—not only did they not understand the science of evolution, they didn't even understand intelligent design. And so here you have people that are supposedly responsible for the education of other people's children, and they have no inkling of what it is that they are enacting. That's truly remarkable."

How important will the Dover case be in coming years? It would be naive to assume that Judge Jones's opinion has settled this battle. Certainly in a country where many believe that Noah's flood created the Grand Canyon and that the Earth is no older than a few thousand years. Dr. Forrest is sure there will be other cases, other opportunities for her expertise in the very near future.

"I guess right now I'm looking to the next occasion when I might have to put what I have learned to use."

PART IX

—

PIONEERS PRESENT
AT THE FUTURE

CHAPTER TWENTY-THREE

JANE GOODALL

Yeti or Bigfoot or Sasquatch. You'll be amazed when I tell you that I'm sure that they exist.

—JANE GOODALL

As a young journalist in 1971 I found myself, totally by accident, in a room with the greatest living archaeologist of the time: Dr. Louis Leakey. There was no one else but we two. I didn't immediately recognize Leakey and did a full, Hollywood double-take and decided there was no way that I could just let this historic moment go by without introducing myself.

But then what? Make small talk? Me, a 22-year-old neophyte journalist? I was so stunned to be breathing the same air he was that I wasn't sure I could put a cogent thought together.

After a few eye-contact moments with the snowy-haired scientist, I found myself walking toward his side of the room, hand extended, and introducing myself as a science reporter. I told him that I had always enjoyed his *National Geographic* television specials and

Jane Goodall, as a guest on Science Friday. *Ira Flatow.*

wished him luck. To my surprise, he brightened, looked straight into my eyes, and said, "Thank you." The door to conversation had opened. Now came the hard part: What to say? I have been star-struck very few times in my career, and this would be the first and longest lasting. I just had to find an icebreaker to open the conversation.

"Dr. Leakey," I croaked, "what would you say is the most important evolutionary advance that sets humans apart from other animals?" That sounded pretty intelligent. I had no idea what he would say: fire making, brain size, transistor radios?

"Oh yes," he began with a twinkle in his eye. "Surely it is precision grip—the opposing thumb and forefingers—that allows us to make tools."

"Of course," I stammered, as I thanked him and retreated back to where the air was easier to breathe.

That brief encounter with Louis Leakey is still one of my most treasured memories, as he died a few years later. And I was eager to share it with a scientist who knew Leaky and had spent many years knowing and working with him: Dr. Jane Goodall.

The story goes that filmmakers working at the Science Museum of Minnesota asked a group of people to name a famous living female scientist. It's not easy to do. Think about it. They scratched their heads and then went to the top of the list, where they found Jane Goodall. Her amazing life story, working with chimpanzees in the jungles of Gombe, Africa, made a perfect subject for an IMAX movie, and so the filmmakers set to work, and *Jane Goodall's Wild Chimpanzees* was the result.

Of course, one doesn't need a pretense to interview Dr. Goodall. She is the preeminent primatologist, the United Nations Messenger of Peace, the founder of the Jane Goodall Institute and the author of many books, including *The Chimpanzees I Love*.

Jane Goodall's early claim to fame was her discovery of just how smart chimpanzees are. She observed chimps making tools—such as fashioning a leaf or a twig—and using it to retrieve hard-to-get food. (They made good use of their precision grip.) The toolmaking abilities of these primates shocked scientists, who had always assumed that only we humans were clever enough to fashion tools.

I had always wanted to interview Jane Goodall and ask her about her discoveries of chimp intelligence. As a college undergraduate, I marveled at the stories my professor told about his hours spent working in the primate lab with chimpanzees, where he was always amazed by how much intelligence and toolmaking ability they exhibited. And I wanted to ask Goodall about Louis Leakey.

INTO AFRICA

Gilbert Grosvenor, chairman of the National Geographic Society, once wrote about her, "She was hardly the image one would project to become an old African hand. Her bush experiences were honed in the genteel English countryside." With that background, how did she wind up in Africa?

"It wasn't exactly genteel," she said. "I wouldn't have described it like that. But animals were my passion from even before I could speak, apparently. I was watching earthworms in my bed when I was one and a half. And I hid for five hours in a henhouse when we had the opportunity to go into the country because I was collecting the eggs and there was the egg. Where was the hole big enough for the egg to come out? Nobody told me, so I hid."

Even at this young age, Goodall would practice a technique that would serve her well the rest of her life: Patiently watching and waiting.

"It was my first wonderful experiment. And then when I was about ten, or eleven, I found the books about Tarzan of the Apes. I fell in love with Tarzan. He's got that wife Jane, so I was terribly jealous of her. And that was when my dream started. When I grew up, I would go to Africa, live with animals, and write books about them. That's how it all began."

Opportunity to fulfill that dream came knocking in the form of an invitation by a friend to stay on a farm in Kenya. "I was working at the time with a documentary film studio in London, which is a great job, didn't pay very much, so I quit that, went home, and worked as a waitress and served people their breakfast, tea, and lunch and dinner till I'd saved up enough money to buy my return fare by boat, because it was cheapest in those days. I was twenty-three and I sort of said bye-bye to family, friends, and country, and off I went on this amazing adventure."

LOUIS LEAKEY

Jane stayed for just a month, not wanting to "sponge off people." Looking to stay in Africa, she found a "boring" job as a secretary that, if nothing else, kept her in earshot of the great explorer of the day: Louis Leakey.

"Somebody said, 'Jane, if you're interested in animals, you must meet Louis.' So I picked up the telephone, cheeky me, and made an appointment to go and see Louis Leakey. He was then curator of the Natural History Museum in Nairobi.

"He was amazing. The first time when I called, to my amazement, he answered the phone, and he said, 'I'm Leakey. What do you want?' It wasn't a very auspicious beginning. But then when I got there, he took me all around. He asked me so many questions about the animals there, and because I had done what my mother said I should do, which was, if you really want something, you work hard, you take advantage of opportunity, and you never give up."

Leakey was impressed with her knowledge. She had been doing a lot of homework learning about Africa and its wildlife, spending many lunch hours in the Natural History Museum. "So he gave me the opportunity to work for him, and he took me with his wife [Mary] and one other young English girl to the now-famous Olduvai Gorge. It was absolutely wild, untouched Africa.

"And typical Louis, there was never any money, so everything was on the shoestring, and the equipment mostly didn't work, and it was a very ramshackle sort of place. And I remember when he first talked to me about going on that expedition, and he said, 'Well, it's going to depend on my wife. If she likes you, you can come.' And can you imagine what it was like when I went to lunch at the house, thinking, *Oh, Dear, what can I do to make Mary like me?*"

Fortunately, she did. Goodall went to work with the Leakeys before they made their famous discovery of the "Zinj" bones in Olduvai Gorge that would change views about the origins of humans in Africa. In fact, she began work in Africa when archaeology was not the exact, rigid science it is today, where excavations are crisscrossed with a grid work of strings carefully marking the exact place objects are removed from the Earth.

"There was no formal digging up a place and marking it on a grid. It was pre- all that, so we just spent all day chipping away in the rock. There wasn't a road there. There wasn't a trail. There was nothing. And all the animals were there, the antelopes, the zebra, the giraffes, and then one evening, there was a rhino, which was a little bit scary, and one evening a young male lion, two years old, totally curious, never seen anything like me."

Goodall had found her calling. She was hooked on a new career.

"When I got there, when I got out to Olduvai, it was like being at home. Louis realized that I was the sort of person he said he'd been looking for about ten years, who didn't care about hairdressing and clothes and parties and boyfriends. You know, I really wanted to be in the wild.

"So he made the suggestion to me. It took him a year to get the money. I mean, who was going to give money to a young girl, a female, who didn't have a degree of any sort, straight out from England? What a ridiculous idea. So I was in England waiting, learning what I could about chimpanzees, while he searched for money and eventually found a wealthy American businessman who said, 'Okay, Louis, here you are. Here's enough money for six months. We'll see how she does.'"

And she did quite well. ⁻

"It was a very, very worrying time because I got to Gombe, again I felt I was at home, but the chimpanzees ran away as soon as they saw me. You know, they're very conservative. They'd not seen a white ape before. And I knew if that six months' money ran out before I'd seen something really exciting, I would have let Louis down. 'Well, we told you so. This is ridiculous.' But fortunately, just before that time came, I saw the first observations of using and making tools, and that was the saving observation, the breakthrough, and he was able to go to the National Geographic Society and persuade them to put some more money in when the first six months ran out. Because, of course, at that time we were defined as man, the toolmaker. That was supposed to differentiate us more than anything else in the rest of the animal kingdom."

Jane's discovery of the toolmaking ability of the chimps would make her famous.

"David Greybeard [the chimpanzee], bless his heart, I saw him crouched over a termite mound, couldn't really see properly. They were still not very relaxed in my presence. I was hiding. But I knew he was using a piece of grass, and a few days later, he and one of the other chimps—I could see them much better—the whole thing, putting in the grass, picking the termites off, picking a leafy twig, and stripping off the leaves, which was the beginning of toolmaking.

"I couldn't actually believe it. I had to see it about four times before I let Louis Leakey know. And then I sent a telegram. And he sent back his famous comment, 'Ha, ha. Now we must redefine *man*, redefine *tool*, or accept chimpanzees as humans.'"

How did Goodall know, when she watched the chimp drawing out the ants on a stem, that she was observing a revolutionary act? As Louis Pasteur observed a century before, chance favors the prepared mind. "I knew because just about two weeks before, I was visited by George Schaller, who'd just finished his mountain gorilla study, and as we sat up on the peak, which was my lookout place from which gradually the chimps got used to me, he said, 'If you see tool using and hunting, those two things will make your study worthwhile,' and within two weeks, I saw them both. It was quite extraordinary. And both times it was David Greybeard."

She named the chimps she was studying, a unique practice that no one else was following. "No, they weren't. And the funny thing was, after a bit, Louis said, 'Jane, you have to get a degree, because otherwise you can't get your own money, and I won't always be around to get money for you.' But he said, 'We don't have time to mess about with a BA, so you'll have to go straight for a PhD.'

"So he managed to persuade Cambridge in England to accept me as a PhD student. And when I got there, it was actually a very unpleasant and hostile reception that I had. I shouldn't have named the chimps. It wasn't scientific. I knew nothing.

"I mean, I couldn't talk about their personalities, these vivid personalities that I by then was beginning to know. I certainly couldn't talk about them being capable of rational thought, which they clearly were. And finally, worst sin of all was that I was ascribing to them emotions, like happiness, sadness, and so forth. But more importantly perhaps, all through my childhood, I had this wonderful teacher, and that was my dog Rusty. So I knew that animals had personalities, minds, and feelings, and of course they needed names. But fortunately by that time I was twenty-seven and I wasn't in it because I wanted a PhD. I was there for Louis."

Besides being ridiculed at Cambridge, Goodall found that her scientific methods and integrity were challenged. "I was even accused of teaching the chimps how to fish for termites, which would have been such a brilliant coup.

"So, the Geographic came in and provided money, and then my late ex-husband was sent out by Geographic, and he got this amazing film, some of which has been blown up for the IMAX, and it's just amazing that the film he took in 1961 has been blown up onto this huge screen. Actually, it's very moving for me to see that. I feel I'm back."

My own earliest memories of Leakey and Goodall were shaped by what I watched on television in the 1950s and 1960s, the *National Geographic* specials. Goodall believes that to be true for many people.

"Whole generations of people saw and were moved by those and got fascinated. And, you know, literally thousands of people have said, 'I'm doing what I do because I grew up with you.'"

David Greybeard, the toolmaking chimp, died from pneumonia in 1958 at the age of around 35, not very old for a chimp. Thinking back to the patience and empathy she showed her chimps, I wondered aloud whether she thought that a primatologist's sex influences how they conduct their work.

"I think, in many cases, it actually does. Louis Leakey always thought women were better as observers. He felt that they were more patient, and certainly it's very often true that women tend to be a bit quieter and more prepared to sit there and let the animal—whatever animal is being studied—tell you things. And it's getting more difficult today because students go out and they have a hypothesis and they've got to prove or disprove it. But in the days when I went out, nobody knew anything, so you just went out there, and everything was new and everything was exciting. It was a tremendous privilege really to be there then.

"And I think that women have a couple of things going for them through evolution. Good mothers had to be patient; otherwise they didn't raise a sufficient number of kids. So if patience can be innate, then the female is likely to have a larger portion of it. And secondly, women have had to be able to very quickly understand the wants and needs of nonverbal beings. That's their own kids. So that too might be helpful when you're trying to learn about another species. And finally, women have traditionally played a role of just being in the family and watching very carefully to see what the relationships are so that [they] can prevent discord within a family before it actually happens."

But can't males share these traits as well?

"Of course males can have them, and there are some absolutely amazing male field study–men doing wonderful field studies. It's just that women seem to be a little more gentle about it."

Scientists like to be surprised. That's why they are in the business they are—to discover the unknown. Goodall was certainly surprised to learn of the toolmaking abilities of our nearest cousins. But an even greater surprise awaited her years later.

"The most surprising and shocking really was when, in 1970—that's after ten years of research—we realized that chimpanzees have a dark side, just like us. I thought they were so like us, but rather nicer. And then to find that they are capable of brutality, that they may even have a series of events not unlike primitive warfare, that they can attack members of another social group so severely that those individuals die as a result of their wounds, and that infants can be killed. And that was very, very shocking."

Why did it take 10 years to discover that?

"Because the boundary patrols are right out at the far end of their range, and I suppose we just weren't following them far enough. But also the war—we called it the four-year war—was a rather specific circumstance. Our main study community divided and the smaller half took up a portion of the range, which they had previously all shared. And when those two groups had separated, the males of the larger group began to systematically annihilate the split-off individuals. It was almost like a civil war. And it was very, very shocking."

It almost sounds like it was a preplanned, well-thought-out tactic.

"Certainly, when they're moving out to the peripheral part of their range, it seems to be planned. Like, one or two males will set off and they'll look back, and very soon, the entire group knows exactly what's happening at that point. The females and young ones usually stop, and they don't go on with the big males."

Goodall gave up real field research in the late 1980s and since then has been in the forefront of animal rights work.

"I was very shocked at a conference in Chicago to see secretly filmed footage of chimpanzees in a medical research lab in cages that were five foot by five foot and totally bleak and barren, isolated, these highly social beings who are so like us in so many ways. And that was really what took me out as an advocate, took me away from pure research, because I felt I owed it to the chimps. They'd taught me so much, they'd given me so much. They really helped to blur the line that people saw as so sharp dividing us from the rest of the animal kingdom. And once that line is seen as blurred, once you're prepared to admit that we're not the only beings with personalities, minds, and feelings, then you have a new respect not only for the chimps, the other great apes, but [also for] other amazing sentient, sapient beings with whom we share the planet."

After decades of watching animals in the wild showing more intelligence than we give them credit for, Goodall has strong feelings about animals used in research. "It was very unfortunate that there was this feeling that it's fine to do anything to an animal, as long as maybe it's for human good instead of saying, as most scientists will, 'Unfortunately, we'll always need some animals.' We've already got alternatives to those. And so I want a mind-set that says it's not really ethical to do this to animals, so let's get together as soon as we can and find ways to do it without using animals. Because, you know, our brains are so amazing. We can do so much."

I wanted to know from Dr. Goodall what was left to be discovered. Were there perhaps any undiscovered large ape species? I was unprepared for her answer.

"You're talking about Yeti or Bigfoot or Sasquatch. You'll be amazed when I tell you that I'm sure that they exist. I've talked to so many Native Americans who've all described the same sounds, two who've seen them. I've probably got about thirty books that have come from different parts of the world, from China, from all over the place. There was a little tiny snippet in the newspaper which says that British scientists have found what they believe to be a Yeti hair, and that the scientists in the Natural History Museum in London couldn't identify it as any known animal."

In this age of DNA analysis, Goodall is hoping that living cells might exist on the sample. "There will be. And I'm sure that's what they've examined and they don't match up." Goodall remains hopeful and confident that if this sample does not yield viable DNA, someday a hair sample *will* be found and prove that Bigfoot really does exist, a belief she has carried for years.

"I'm a romantic, so I always wanted them to exist. There are people looking. There's a very ardent group in Russia, and they have published a whole lot of stuff about what they've seen. Of course, the big criticism of all this is where is the body? You know, why isn't there a body? And I can't answer that. And maybe they don't exist, but I want them to."

Besides its use in finding Yeti, Goodall says DNA technology has changed primate research.

"That's been very exciting because the one thing we never knew for sure, although sometimes we could guess, is which male fathered which infant. And with DNA profiling techniques, which can now be done from fecal samples—you don't even need hairs—we now are beginning to identify the fathers. That means that we can look at the relationship between a particular adult male and an infant and find out if there's any special behaviors which seem to indicate that in some way they know. Now we don't know yet, but it's fascinating. Sometimes our guess is absolutely confirmed. We found an example of incest, which is very rare. So it's a fascinating new field for us.

"One of the most fascinating areas for research is cultural differences between different populations across Africa or even different neighboring communities. And of course, it's still controversial as to whether chimpanzees can have culture, but I define it very simply as behavior that's passed from one generation to the next through observation, imitation, and practice. And tool-using behaviors differ quite markedly across the species range in Africa. Sometimes it's due to different environments, but very often it seems to be due to the young ones seeing what the older ones do.

"Now we've just begun to skim the surface of these differences, but even as you and I are speaking, chimpanzees, along with their cultures, are being wiped out right across Africa. From about two million a hundred years ago, to the very maximum two hundred thousand today. And that's more likely to be one hundred fifty thousand spread over twenty-one countries, mostly in tiny, isolated fragments where there's no possibility for long-term survival because the gene pools aren't big enough."

And they're dying out why?

"They're dying because of habitat destruction, as human populations grow. They're caught in wire snares set for other animals, but they catch the chimps and gorillas, for that matter, and they either die of gangrene or lose a hand or foot and can't compete very well reproductively. But the worst threat for chimps today is the commercial bush meat trade, and that is the hunting of animals for sale in the big towns. Not subsistence hunting, which has gone on for hundreds and hundreds of years, but this has happened because the logging companies have made roads into the heart of the last great forests of the Congo Basin. Hunters go along the trail, they now have transport, they shoot everything, they load it on the truck, they take it to the towns, where the elite will pay more for it than chicken or goat.

"And it's not sustainable at all. And it gets worse because you've got your logging camps, two thousand people or more, the loggers and their families, who weren't there before. And now the Pygmies, the indigenous people, are paid and given weapons and ammunition to shoot for the logging camps, and that's not sustainable either. The logging camp moves, the Pygmies have had it. They've lost their culture. They may have trees standing from sustainable logging, but they'll be dead forests."

Don't the indigenous peoples understand the cycle? "There's nothing they can say about it. They don't want the loggers to come in any more than the people in Ecuador want the oil pipelines to come in, but what can they do when big-business interests are put before the interests of the people living there?"

So it's just a short time before we lose the chimps. "We're working very hard to do something about it. The United States government, through the State Department, has put quite a large sum of money into a Congo Basin project. And best of all, President Bongo of Gabon, which has the largest area of unlogged forests, has just taken twelve areas away from logging concessions."

Goodall is also worried about the state of the environment closer to home. She says that the fight against terrorism is threatening to overshadow environmental concerns in the United States. "Well, it has. It does. Look at the drilling for oil in the Alaskan Wildlife Refuge. It was blocked, but now probably it's going to be going ahead although in somewhat modified form. But directly after the eleventh of September [2001], it was very clear to me as I traveled around the country that people were reluctant to admit that they cared about the environment in case they would seem unpatriotic. And fortunately, gradually, people are coming out of that mind-set because, you know, if we let the planet continue to deteriorate, we really are in a very, very bad state. And if we continue to let that happen, then the terrorists finally will win, because for our great-grandchildren there will be nothing left."

Celebrities tend to get noticed; they draw the media. So does Jane Goodall now see herself as an effective spokesperson for the environment?

"I do spend a lot of time talking to young people, but also people from all walks of life and all ages. And one of the remarks that's so often said to me after a lecture, people come up and they say, 'You have reinspired me to do my bit. You have made me feel that my own life is more worthwhile. I feel that I've been just sitting doing nothing. Now I want to do what I can.' So until there's a groundswell of people prepared to accept the tough decisions that may affect their purse to some extent, then we'll never get the right legislation.

"I do know that when talking to people who perhaps think very differently, the only chance you have of getting them to think in a different way is to touch the heart. And if you're strident, if you start accusing people, if you point fingers, then you immediately see the eyes glaze over and you know that you're not getting across. I think that so much of what goes on that, in my view anyway, is a mistake is due not to any kind of criminal intent but simply because people honestly haven't understood. So I feel that that's my job. My job is to help people understand and to think about the future. I mean, just imagine what this world would be like if we went back to the old tradition of the Native Americans, who said every major decision has to be made with the question 'How will this affect our people seven generations on?' [in mind]. Even if we could just say two generations on, even one generation on, it would be helpful."

CHAPTER TWENTY-FOUR

IAN WILMUT: DOLLY PLUS TEN

Meat and milk from clones of adult cattle, pigs and goats, and their offspring, are as safe to eat as food from conventionally bred animals.

—U.S. FOOD AND DRUG ADMIN-
ISTRATION, DRAFT REPORT,
DECEMBER 2006

In 1996, a fetching newborn lamb named Dolly made headlines and magazine covers around the world. She looked nothing like the ewe that gave birth to her, because Dolly was the first animal cloned from an adult cell.

Dolly went on to mate and give birth to six healthy lambs. Though she suffered from illnesses, such as arthritis and lung disease, normally associated with much older sheep, she was such an amazing accomplishment that when she died in 2003, many people around the world who had never seen her in person felt that loss.

Today, of course, animal cloning is almost common. But Dolly's creation triggered fiery debates over human cloning that are still very much alive.

Ian Wilmut is an embryologist and professor of reproductive

science at the University of Edinburgh in Scotland. He was a member of the team at Scotland's Roslin Institute that cloned Dolly back in 1996. He is opposed to cloning human babies, but he says he's a passionate advocate of what he calls a restricted form of human cloning.

"It's amazing to think that it's already more than a decade ago that she was born. And a lot has happened, but I think there's a lot still to come. Using this technology to try to understand human disease, to develop new treatments for them." That's one of the immediate benefits, says Wilmut. Looking farther into the future, perhaps "being able to use it to stop the birth of children with inherited disease. I think there are a lot of things which I would encourage people to think about and to be optimistic about this technology and what it can offer."

Now that scientists have successfully cloned a variety of animals, people are beginning to talk about the possibility of cloning people. After all, it's not unreasonable to expect that parents might desire to have a clone of a deceased young child, tragically lost to them. But Wilmut remains cautious.

"The application that I've never supported is the idea of producing somebody who is a genetically identical twin of a person who is here. To me, it would be acceptable to produce embryos from which you could get cells and to produce a baby in which you had corrected a genetic disease. But it would not be a clone of somebody who was here already. It would be, if you like, a clone of an embryo. And it's being done for a different purpose. So it's very important all the time to say exactly what you have in mind."

Is it possible that there might be a rogue scientist somewhere in the world who might attempt to clone a human being? Some have claimed to be able to do so. . . .

"I think it is extraordinarily unlikely. The people who've talked about this in the past have been advertising themselves but actually not doing anything. And I think this is a fear which is greatly exaggerated."

Would it be possible to create a united, international front against cloning people?

"I think that would be preferable. But given the different cultural histories, it may not be easy to achieve it. And so I'd be quite content to see each country preparing its own regulations. And it is, of course, very disappointing that [the United States] hasn't done it. The United Kingdom, where I work, has some of the most liberal approaches to this. But it is very tightly regulated, which I think is entirely appropriate."

But how can regulation stop someone with the knowledge and tools who desires to ignore the boundaries of accepted behavior? Wilmut doesn't believe it inevitable that some scientist will create a human clone. "No. It might happen. But I think it depends on how we behave. It is, of course, possible for societies to come to different conclusions. But a perception that we should have is that there are disadvantages to most technologies. What's key all the time is for societies to discuss things, to be informed about, and then to prepare regulations."

EXPECTING TOO MUCH

Sometimes, technologies are oversold—that is, when they first break into the news, we hype them as saviors, as disease cures to the point that when they don't produce results quickly enough, we think of them as failures. Has cloning been overhyped?

"I think in the short term, perhaps it has. There are different time scales. People doing research have to be optimists to follow that way of life. And it may be that inadvertently we do oversell things. But in the long run, no, I don't think it's been oversold. There'll be a lot of things [that] come from this area of research.

"There are people who out of religious beliefs regard an embryo as a human being. Because of that, they can't morally condone the use of embryos in basic research, in cloning, or embryonic stem cell research.

"I understand that to some people what we're suggesting is deeply offensive. And of course, I accept that there are different points of view. In terms of working with embryos, I would wish to be sure that they understood exactly the stage of development that we would be working with. The embryo is smaller than a grain of sand. The micromanipulations we carry out with very sophisticated equipment and very skilled colleagues [occur] weeks before the time when there would be the beginnings of a nervous system.

"To me personally, the key human characteristic is being conscious, aware, being able to form intellectual relationships with people, something which, of course, would happen much, much later in the life of a developing person. But until the point when that begins to occur, which is sometime after the stage we would take cells, I think this is a *potential* person but not yet a person. And that's the reason why, to me, it's an acceptable thing to do."

Wilmut's work has been with animals. And he believes lots of advances are still to come. Probably not as dramatic as cloning humans but still quite worthy of scientific research. For example, take cows. People take cows for granted. "But let's consider instead that it is possible to change cows so that they could not be susceptible to mad cow disease. It may be possible to change cows so that they would be resistant to foot-and-mouth disease. Now, I would suggest to you that those are two objectives which it is well worth considering not only for the benefit of the people who may ultimately consume milk and meat from them but also for [the benefit of] the animals themselves."

And what if we could get farm animals to not only supply us with meat and dairy products but to also be able to grow spare body parts—hearts, kidneys, lungs—for us too? "One project, which a couple of groups certainly are still pursuing intensively, is the idea of changing pigs so that their organs can be used in people. And there is at least a handful of papers now showing that by making a particular genetic change, the organs are able to cope with blood from primates, not from human beings, but much better than anything in the past. I think it's still some way away. There are a lot of challenges still to overcome, but it is something that people are still considering."

CLONING PETS

Okay, so forget cloning whole people and making body parts from animals for a while. What if you want something more basic? What about the ability to clone Fido? What if you want to "bring back" your favorite pet? It's already been done with dogs and cats. But the prices have ranged into the tens of thousands of dollars. When might that price come down?

"It's very difficult to know all of these things, I'm afraid. With the present technologies, as you know, the efficiency is very low and the cost very high. What we need for many of these applications is another jump in efficiency, which is comparable to that which enabled us to produce Dolly. Whether it will come from one more big step or from lots of little ones, which will accumulate, it's impossible to know. And who knows what it will be or when it will come. I'm sorry not to be clearer than that."

One of the greatest hurdles to cloning is how inefficient it is. It takes many tries, sometimes hundreds and hundreds of tries, to clone an animal. Understanding why that is, says Wilmut, is one of the challenges of the future. "And that would then facilitate everything that we do. I think a fairly near thing is to be able to produce cells that have the characteristics of a patient with ALS [amyotrophic lateral sclerosis, also known as Lou Gehrig's disease], for example, which is a relentlessly progressive disease, which means people lose control of their limbs and ultimately die because they stop breathing. A very sad situation. It isn't understood."

Wilmut points out that one of the problems in learning how to treat ALS is that the affected nerves lie deep inside the body, "so the researchers can't get access to them to study them. If we produce cells from cloned embryos from a person who had inherited the disease, it would be possible for the first time to study those cells as they begin to change for the very first time, to understand it and hopefully to develop a high-throughput drug screening system that would enable us to assess thousands of drugs and hopefully identify something that could stabilize the position of patients. I think it would be five, ten, fifteen years before that program would be completed. Probably ten years before it would be completed. But I think there is a realistic hope that that would bring forward the first drug to be able to stabilize the position of patients. And just imagine what that would mean to somebody."

And once there is such a breakthrough in the treatment of a major disease, opposition to embryonic research would melt away.

"I believe that. And we haven't got time to list the number of diseases which could be studied in that way. I mean, they vary from some causes of sudden death because of heart failure, cancer, psychiatric disease, other neurodegenerative disease. There's a huge range of inherited diseases that we don't understand and for which there isn't a treatment. And, you know, among your audience, there will be many people who already have one of these conditions or will have a family member or a work colleague who has these conditions. And because we're living longer, we have more chance of succumbing to them and suffering that fate ourselves.

"I mean, I do respect the opinions of other people, of course. It's their right to have different views. But I think it will become acceptable."

So how does this vision drive Wilmut's future? It sends him back to the laboratory to understand cloning "because there's another thing which I believe ultimately will be the greatest inheritance of the Dolly experiment, which is by understanding cloning. I think we may be able to change cells of one type into another type, without making an embryo. And that would facilitate the idea of patient-specific cells."

Patient-specific cells are highly desirable. It means that the body does not reject the cloned cells, since they come from the patient's own body.

"It's by making biologists think differently [that] the Dolly experiment has been most important of all."

CHAPTER TWENTY-FIVE

THE WIZARD OF WOZ

My idea was never to sell anything. It was really to give it out.
—STEVE WOZNIAK, COFOUNDER,
APPLE COMPUTER

In the 1970s, a group of engineers, hackers, and computer enthusiasts met in what today we call Silicon Valley. It was a regular meeting of the Homebrew Computer Club: a discussion of chips, wires, circuits, and general nerdy tech-talk. (The word *geek* had yet to be invented.)

But at one meeting in 1975, a club member named Steve Wozniak set up a computer he had designed and built on his own. It had features that no one had ever seen before, features such as a keyboard. You could type on it instead of having to flip switches or shuffle punch cards. It had a TV screen, with numbers on a screen. It sparked a revolution in computing, putting the *personal* in *personal computers.*

Wozniak, or "the Woz," as he's affectionately known by his

followers, went on in 1976 to cofound Apple Computer with Steve Jobs. He became the driving engineering force behind the company's innovative early designs. But he'll never forget those club meetings.

"The Homebrew Computer Club was the highlight of my life. Every two weeks, we'd meet on a Wednesday night. I was too shy to ever talk in the club meeting, but the way that I could communicate sometimes was by doing good designs. I was very skilled at a certain type of circuit design."

Wozniak was broke. He had no money. Had no savings account. "I had to pay cash for my apartment because the [rent] check had bounced enough times." But lack of liquidity did not deter him from following his dream: building a new kind of computer. Back in the 1970s, computers were large, room-filling affairs called mainframes. They had no keyboards for word processing or spreadsheets. Those tools didn't even exist yet. What they did have was a bank of punch-card readers. You collected your data and transferred it to punch cards, stiff paper ballots with holes mechanically punched in them. Stacked in a deck, they were fed by hand into the mainframe.

Wozniak wanted to create something totally new: a typewriter input and a TV screen display. He tried out this idea on his club companions.

"I would bring down my color TV set, a Sears TV with a cable snaked into it, and hook it up to the circuit of very few chips and then a little keyboard you could type on. And I was trying to impress people with 'How did he do it with fewer chips than anyone could ever imagine?' It wasn't really to show the world 'here is the direction the world should go.' It was to boast, to be clever, to get acknowledgment for having designed a very inexpensive computer."

Wozniak was no businessman. He created his computer purely for bragging rights. "My idea was never to sell anything. It was really to give it out. So I started passing out the schematics and the code listings for that computer, telling everyone, 'Here it is. It's small, it's simple, it's inexpensive. Build your own.'"

All of that changed at the hands of the other Steve. "Steve Jobs came along. He had more of the future vision: 'We can bring this to everyone; we can start a company; we can sell it.'"

Jobs and Wozniak were a perfect match. The entrepreneur and the geek, one complementing the talents of the other. "When you're designing and inventing the way I did, every minute of your life, every neuron in your brain into trying to think about the little [computer] code and how you can maybe have one less line of code and a little bit more straightforward from the beginning to the answer. And you don't have time to think about companies and products and how would I build this. So Steve Jobs and I were a very necessary pair and his first idea was really not even to build a computer."

Jobs, says Wozniak, came from a world of surplus computer parts where you could go into a store crammed with old cardboard boxes full of electronic parts and walk out with armloads full of gear costing next to nothing. (I knew this world very well, having spent countless hours of my youth in such stores in New York City.)

Jobs's idea, says Wozniak, was to build on what he saw in these stores: assemble the parts into computer components and sell them at a small profit. "He said, 'Let's sell PC boards for forty dollars. We'll build them for twenty dollars and sell them for forty dollars.' Neither one of us could be sure we'd get our money back on this investment, but we just wanted to have a company of our own for once because we were best friends."

ONCE A GEEK, ALWAYS A GEEK

Wozniak is the epitome of the *geek*, a nom de guerre he wholeheartedly embraces. "Being an electronic genius was a reputation I had," focusing so much on math and science that he had no desire to be involved in the other normal parts of the world.

"It's really more a characteristic where you don't socialize. You don't talk the normal languages. You kind of feel embarrassed. You're an outsider. You become very scared to open your mouth around normal people. You hear people coming up, doing their talk about 'Hi, nice day,' and the small talk starts up, and you don't even know the clues of how to do it. I don't to this day."

When Wozniak started Apple, he had no college degree. They didn't give out degrees in a field that interested him; personal computer design did not exist yet. Woz would create it. He designed the Apple I and Apple II computers, radical enough for their time. But when Kaypro, Sinclair, IBM, and the other computer makers imitated the Apple II keyboard–screen idea, he and Apple moved on to adopt an even more radical approach to computing that Jobs had seen during a visit to the Xerox Palo Alto Research Center: the computer mouse and the graphical "point and click" user interface.

Apple's version would be called the Macintosh, and Wozniak was quick to join the Mac's design team. "My best friends in the company, which are usually not the high-up guys, the ones accorded with the top degrees, but the ones who are the interesting people, the ones who never went to college but can design things with almost no parts and no waste the way I did, and write the cleverest little code and solve any problem and just loved what they were doing. They were in the Mac group, so I joined the Macintosh group."

So Wozniak remains as much a geek today as he was in the 1970s. He's just a few bucks richer. With little prodding, he will eagerly show you his giant digital wristwatch made of old numerical light-bulb "Nixie Tubes." He'll launch into a free-ranging, free-associating monologue about how much he loves what he does.

"I had played this game so long that I had all these little tricks in my head that I can't even explain. My head carried all sorts of circuits." Like a mental Rubik's Cube, Wozniak could twist the designs around in his head first one way, then another, looking at all the angles, all the possibilities, using different parts of the chips that he had amassed for free. "Nothing was wasted; absolutely zero waste. I told this story recently to the Resource Recovery Association—recycling—and they loved to hear I didn't believe in waste."

As is customary for a geek, Wozniak, an introvert nonetheless, is very proud of his breakthrough achievements in computer design. "I really believe I know why my designs were better than any other human being's," he says, casually comparing his superior software design technique with the masters—"like Mozart would do it." Yet even this computer maestro, the man who revolutionized personal computers forever, did not see the next great revolution, a tsunami called the iPod, the brainchild of his old partner and friend Steve Jobs.

"I saw lots of music devices. I loved playing with music devices. And like most of the world, I thought of a music device as a music device. Steve Jobs tends to look beyond that, and he doesn't see a music device as having any importance at all—how fast it is, how many songs it can hold. He sees music itself, to a person, as being the important thing."

Jobs, he says, saw that merely supplying songs to people was not enough. Just as Edison saw that a lightbulb needed a system of electrical generation, wiring, and infrastructure to be successful, Jobs saw that a song could not exist by itself. "It took a service that could sell you songs at a reasonable price, download them easily with almost no steps, no work, no hassle on your part." You would put it into a computer and then plug the iPod in, "and magically, with no steps on your own, it gets into this device you can carry with you, a satellite of a computer."

What special talent does Jobs have? Why was he able to intuitively foresee the popularity of the iPod when no one else did? Wozniak has seen this uncanny ability in his friend before. "All through time in Apple products, even from our very first ones, that's how he looked at the world: You don't really want a piece of technology, a certain type of chip. What you want is a solution to a problem in life, some cause, some issue that you want in your life that'll help you. And it's how do you make that almost one step—say it and it happens."

As for himself, Wozniak is the first person to tell you that he lacks the ability to see such innovations or predict the next big thing. "I have to be honest. I do not like to talk about the future. I don't like to be one of those people. It's so easy to have a very vague idea and say, 'Oh, computers will be three-D-ish' and then ten years later I'll say I predicted it ten years ahead. I don't think that's honest, and I don't think that's valid and worth anything. Predictions can sound really good if you're good with words and can express them eloquently and give people ideas and inspiration in their head. But I'm not really good at that, so I don't want to."

I'M A MAC; I'M A PC

Wozniak recalls the early days of competition between Microsoft and Apple over which company's computers would dominate the personal computer market. Having captured 90 percent of the market today, Microsoft is universally seen as the victor. But Macintosh users are still very loyal and very vocal about the superiority of the Mac over the PC. Wozniak recognizes that many Macintosh users harbor bad feelings toward Microsoft, but he says, "I never sensed really bad blood between Microsoft and Apple. It comes from the fact that there were points in time when we were told that because the Macintosh had nice little pictures and icons and a mouse and cursors and menus—it had all these wonderful things—and in the PC you had to type and memorize commands to move a file from here to there, we were told by the businesspeople, 'You're a toy.' We didn't like that. And we knew we had a better machine just as capable of any calculation, and yet it was being called a toy. That made us feel we were being put down for something that we knew as users was a wrong assessment. And now of course every computer in the world is a Macintosh, so we were right. Nobody ever came back and apologized and said, 'Hey, you were right.'"

What Wozniak means, of course, is that virtually all computers have adopted the Macintosh's pioneering use of the mouse "point and click" system, without admitting the superiority of its design. Have you ever heard Bill Gates say, "Thanks, Steve!" So, is iPod Apple's revenge?

"No. Apple's revenge is just the fact that Windows—PCs—all became Macintoshes in a way."

As for the Macintosh zealots, "I call them Macintosh bigots a little. They say, 'Oh, no, only the Macintosh is the good one.' I don't like to be that way."

APPLE: THE BANE OF MY LIFE

Wozniak views the fame and fortune Apple Computer has given him as an impediment to who he wants to be. He says the computer company is the bane of his existence. "I want to be Steve Wozniak, who I decided I was at a young age, and not change. I want to go back to school and get my college degree like I would have without Apple." Which he eventually did.

"I want to teach young kids like I would have without Apple." Which he did too, completing a lifelong dream. "I wanted to be an elementary-school teacher my whole life. In sixth grade I told my dad that I wanted to be an engineer first, like he was. But secondarily I wanted to be a fifth-grade teacher because my teacher was so important to me and was giving me the education that was going to take me through life and through this world. And all my life that thought was so important: *I want to get back to education.*

"When I was in college I paid attention to child psychology portions of our psychology classes. I watched other people work with babies. And I saw the baby as developing like a computer, and it intrigued me. I wanted to do that.

"Young children were always so important to me. Adults should treat children with more respect. We should put more monies in our schools. I grew up on that side of the coin." His chance to help kids finally arrived. "I had a lot of money. I had much more money than you ever need in your life to live on." So Wozniak donated computer labs to school districts. But his philanthropy did not quite scratch his itch to give.

"So I decided you should really give yourself. I went and I started teaching computers to young kids, to fifth-graders at first, later to sixth-, seventh-, eighth-, and ninth-graders. I also started teaching teachers. And that was back in the days when we'd wire up the labs ourselves."

After a long and successful career that includes receiving two honorary doctorates, induction into the National Inventors Hall of Fame, and being awarded the National Medal of Honor, the Woz regards his sense of humor as one of his most prized possessions.

"I've been having a lot of fun every day. You know, pranks, jokes. But it actually started with a lifetime philosophy." Wozniak came to the realization at the early age of 20 "that it was less good to be successful and better to have a laughing life. Laugh more than you frown all through your life. Because on the day you die, which one would you have said had the happier life, the better life? And so I put a lot of humor in my life.

"I just believe [that] in whatever you're going to do, even if it's work, have a little bit of fun attitude about it. You can be happy."

PART X

—

THE ULTIMATE
COMPUTER

CHAPTER TWENTY-SIX

WHITHER CYBERSPACE?

They say a year in the Internet business is like a dog year, equivalent to seven years in a regular person's life. In other words, it's evolving fast and faster.

—VINTON CERF

In the mid-1990s, very few people had heard of the Internet. I couldn't get my publisher to let me write a book about the Internet. "What's that? Who cares?" (They are always *so* up to date on the latest techno trends.) Today, there are more Internet-related books, articles, and blogs than you can shake a stick or point a mouse at. So predicting the future of the Internet is harder than reading a cloudy crystal ball. The Web is a world in a constant state of flux, from issues of legal control and governmental influence to the basic networking that holds the Internet together. Predicting the direction of the Internet is as risky as predicting the future of computing a generation ago. Some of the world's most important technologists, from Bill Gates on down, were so wrong about the future of personal computing—and the Internet—that looking into our own crystal ball can be quite risky.

But no matter what "next best thing" comes along, some basic questions remain the same:

- Can the network continue to grow to take on new technologies? We've got telephones that talk over the Internet. We've music and video iPods. Who knows what Steve Jobs has in store for us? Will the information superhighway expand fast enough to meet the additional traffic all the new goodies and gadgets will require?
- Is the Internet of the future a network like an electrical utility, where you pay for the amount of stuff that gets sent to you? Or is it like a cable service where your service provider controls the level and quality of your service, depending on how much you pay?
- How can we be assured that the Internet remains secure? That people will not be eavesdropping on our Internet calls and messages?

As it's almost silly to look too far into the future, let's look at what the Internet may look like by about 2017. Technologists, lawyers, and consumer advocates are already thinking in this time frame. Will people be doing things on the Internet similar to the things they are doing now? Communications is still the number one use of the Internet, says John Horrigan, associate director for research at the Pew Internet & American Life Project in Washington, D.C. Horrigan studies how people use the Internet.

"People still gravitate toward e-mail and IM-ing [instant messaging] and those sorts of things. Following closely are Internet searches. People scratch their informational itches every day in a variety of ways by [going] online and doing a search. And we find, for those of you interested in the media business, that people go online increasingly for news and leave behind other types of media such as national TV newscasts and local newspaper readings. So lots more communications activities, particularly richly interactive types of applications, are popular among Internet users. As bandwidth increases, people are more comfortable in watching videos online and taking advantage of some of the entertainment applications."

STRIKE UP THE BANDWIDTH

And there is plenty of room to grow. That bandwidth—the number of lanes on our info superhighway—is already high and underused. The basic structure of the Internet—its backbone of high-speed fiber optics—is not where the Internet's limitations lie. The backbone is not the problem, says Tim Wu, professor of law at Columbia University and coauthor of the book *Who Controls the Internet? Illusions of a Borderless World.*

"There's a huge amount of capacity. Back in the 1990s, companies, even like Enron, built massive amounts of capacity into the backbone. The real bottleneck, and this has been true since about 1913, is the pipes and the wire that goes to people's homes. Right now, we have the same networks—the cable and the telephone networks—that we've had since the 1960s. They've been updated a bit, but basically it's the same old stuff, and the challenge over the next decade is whether we will really get the same kind of speeds on the backbone in the last mile, which is to say right to people's homes." Think about it. Do you have a high-speed fiber optics cable running into your living room? There *is* one on the telephone pole outside your home. But that's where it ends. Bringing that "fiber" the last 50 yards into your home, from your curb to your couch, will be the next great high-tech reason to celebrate.

"There are great incentives to upgrade the network simply because that's what users want," says Horrigan. "When we look at our surveys at the Pew Internet Project, most people who are signing on to broadband—and today forty-two percent of Americans have high-speed connections at home—sign on for the speed. They don't tell us that 'the price fell and I decided to switch.' The longer people stay online, the more they want to do online, and their demand for speed increases as they get more experience on the Internet."

And what do they want to do with all this speed? Surprisingly, it's not just to play video games. It's what Horrigan calls "user-generated content."

"People like the speed in order to post things online about their lives, about what they're doing. They like to post online their own creativity. Some of that stuff is not going to be widely interesting to the world at large, but it's interesting to users and their social networks. So they're going to want the entertainment services, but they're going to want to have an Internet that enables them to express themselves online."

NETWORK NEUTRALITY

Yes, this generation of Internet users is all about "me." My photos. My blogs. My videos. My baby pictures. Look—and watch—me. But who will control who gets to see "me"? Wu wonders what your Internet provider of the future will look and feel like. He coined the phrase *network neutrality* to describe the issue and worries whether the Internet will remain the wild and woolly, freely open space it is today, where all Internet users are treated equally.

"There's a basic philosophical divide between two approaches, one which is more like cable, which is basically what we've had since the '70s, which is a centralized entity like the phone company or the cable company, and some producers create it. And the kind of world where people just throw up random things, and that's what people watch. How you build the network, how the network gets built out will determine what kind of culture prevails."

Will it be something like what the Internet is now, a folksy kind of culture where people watch really bad quality videos of themselves because they think it's funny or they read blogs almost as much as they'll read the *New York Times*? Or will it be something more like the "old media" world, where the content is controlled by centralized decision makers? A place where you get to choose from a limited menu, as you do on cable television, and have to pay extra for "premium" services, as you do for premium cable channels such as HBO.

"That's the philosophical battle behind net neutrality," says Wu. In other words, will the network remain neutral—equal—for all players, or will some parties have priority over it and make you pay extra for some services? "The Internet today is very decentralized. People come up with stuff, whether it's on YouTube (where you can broadcast yourself), whether it's eBay or whether it's people's blogs." The phone companies, who are increasingly becoming Internet providers, want to limit that freedom, says Wu. For example, they don't like Google "free-riding on our pipes." In effect, Wu says, they are saying, "We want to create premium, a tiered Internet. Maybe we want to start choosing more of what the user's experience is like. Start tailoring it more, choosing one company over another. And there's a big reaction to that among users, and mostly negative, because they like the world where, you know, a blog like *Boing Boing* can be as popular as CNN, and it's four people who are just goofing around part time. And I think Americans like that folk-culture feeling from the Internet. They don't really like big media, media consolidation. It's one issue that has a lot of grassroots support." So much so that Congress has been looking at various bills that would ensure net neutrality.

"I hope that it retains its character as a decentralized medium that basically everyone uses, that just has a lot more bandwidth."

Perhaps the greatest challenge to the Internet of the future is keeping the trust that people put in it. The Internet was built on trust, defying the cynics at every step.

"People have been saying since the '90s that so many things wouldn't happen on the Internet because no one trusts the Internet," says Wu. "It's been a bit of surprise over the last decade of how many things have happened on the Internet, even though it's not as secure as some people might like. Look at the progress that has been made, purely on the basis of trust. Take Pierre Omidyar, who founded eBay.

His whole founding principle was that people trust each other. And eBay has never been all that secure, but it's managed to be successful nonetheless." And now we have blogs and podcasts and other personal services that may not be very secure.

As for the future? "I hope it looks like it's a place that you can securely do all sorts of online transactions," says Larry Peterson, chair of the Department of Computer Science at Princeton University in Princeton, New Jersey. He's the director of the PlanetLab Consortium, an experiment in new networking technologies. Looking forward, Internet security will remain at the top of his priority list.

"That's something that people are really worried about, between the phishing [attempts to steal your identity] and hijacking of connections and so on. We may be reaching a point that the average user loses trust in the Internet, and that's clearly something we have to pay attention to over the next ten years."

This is a widespread problem, he says. It's not just your obvious, secure online bank account, your Social Security number, or your Internet passwords; online content of all kinds is under attack, all the time. There are ways to listen in on your Internet voice conversations, even the encrypted ones. "It's clearly an issue that we have to start paying attention to."

"The world is a very different place than it was thirty years ago," says Horrigan. "The original assumption of the Internet was that everyone was a good guy and [that] if there were bad guys, they were on the outside. And all I had to do was create a safe world on the inside, and we all trusted each other and everything was fine.

"But of course that's not the case today. The adversaries aren't just on the outside; they're everywhere. And so we have to rethink who we trust and how we build a network up based upon those trust relationships."

KEEPING IT OPEN AND DECENTRALIZED

"Despite all the security risks, the Internet has grown beyond expectations," says Peterson. "But on the other hand, what I see happening is what I would call the Internet fragmenting into gated communities. This is almost like the neutrality aspect of the issue, that there are places you just can't get to and the universal connectivity of the original Internet is deteriorating."

Internet experts call this the "Balkanizing effect," the creation of destinations on the Internet that are off-limits to others. One prime example, says Wu, is China. "The Chinese government, to a degree which wasn't true in the early Internet, has imposed the level of watchfulness and control over the Internet which is somewhat unprecedented." Balkanization is often at the country level. "Countries have different ideas of what they would like the Internet to be," and so the Chinese seek to control what their citizens can see on the Web or where they can surf. Jonathan Zittrain and Benjamin Edelman of Harvard University Law School tested the Internet filtering capabilities of the Chinese government in 2002 and found almost 20,000 Web sites that were accessible from the United States but were not accessible from China. The sites ranged from those of the American Cancer Society and the MIT Alumni Association to the *Irish Chronicle* site and the official Web site of the state of Mississippi.

Wu does not quite see Balkanization as a threat but rather as a phenomenon, particularly at the country level. "It's not that countries won't be able to talk to each other; it's just that they will talk to each other less. When the Internet was much smaller a decade ago, there were only so many people on it, and so everybody always talked to each other. It's becoming more and more of a national medium. So the Internet in Germany and in France is in German or in French," or in China it's in Chinese and in Japan it's in Japanese. It just takes on more national characteristics.

"It's not necessarily bad. Japanese people love using the Internet from their strange-looking cell phones," Wu says. "Chinese people, for some reason, love chat rooms. Americans love blogs. They just like saying what they want to say. And so the different countries are kind of shaping the Internet to their own culture. It used to be a medium that was floating in outer space, and [when] you went there, you became this netizen." Now it's becoming part of national cultures. "And I don't necessarily think that's a crisis. I think that's kind of natural, and I think that over the [next] 10 years, we'll see more of that."

REDESIGN?

The Internet is facing challenges that its original designers never dreamed of, says Horrigan, like the mobile devices that didn't exist decades ago at the dawn of the Internet. "Originally the Internet assumed that the things that connected to it, the computers, were always plugged in, they never move. And clearly none of that's true today, because we are a very mobile world and we have mobile devices and we have very small embedded devices that are going to be connected to the Internet very soon." In addition, the rapid growth of video-sharing sites, such as YouTube, places great stresses on the Internet never envisioned by its inventors. "It's these sorts of assumptions that are starting to break down and causing people to think if we stood back and had a chance to design this thing over again, how would we do it?"

One of the strengths of the Internet design, says Wu, is that it really hasn't been optimized for anything. "When it began it was used for e-mail and bulletin boards, and the World Wide Web grew on top of it, and then instant messaging grew on top of it, chat rooms grew on top of it, Google. All these things grew and grew and grew, and even phone service has been replicated. So that original design turned out to be a lot sturdier than people thought. And the question is whether it's possible to improve on it."

"The Internet was something that we were able to tinker with and innovate with at one point, and it's now such a commercial success that we aren't able to do that," says Horrigan, speaking for Internet designers and researchers. "So we naturally gravitate toward the places that we can innovate, which is on top of the Internet."

And there are a lot of people who believe that the Internet is just fine and that we will just go off and build new services on top of it. But the real debate gets down to can we, in fact, solve all the problems that we foresee by only working on top of today's Internet? Or do we really have to reconsider how it's designed at the core?

One of those problems that seems to never go away is ability to use the structure of the Internet's network of computers to launch cyber attacks—to attack someone's computer and bring it down. Take a recent cyber attack on CNN.com, says Peterson, where hackers "commandeered a thousand 'zombies,' machines that were infiltrated around the world, and were made to start sending traffic to CNN.com. Such a load was put on their system that no one else could get to it. The CNN Web site and its services and employees were effectively shut down. This is a common cyber attack called 'denial of service.'

"That problem is an impossible one to make go away by building on top of the Internet, because there's already access to the underside of the Internet [so] that you can still send those packets. And so if all you can provide [is to] enhance services on top, it's very difficult to correct something that's inside," says Peterson.

"There's always this debate," says Wu, "as to whether things can be improved on top of the Internet or whether you have to rip up the highway and fix it that way." Wu says it's not wise to rip it up and start over. The Internet's original, very simple design has a lot of utility left in it. "I'll give you one example. When I worked in the telecom industry, no one ever thought the Internet would be useful for phone service, for dialing people up. Everyone thought, *It's just a lousy design; it'll never be useful.* But with the right amount of bandwidth, companies like Skype and other [Voice over Internet Protocol] companies have been very successful. It's really surprised a lot of engineers.

"Sometimes it's very hard to improve on simplicity. I'm not a security expert, and I think maybe security [like cyber attacks] is one area where it's hard [to improve]. But it has been surprising: video, voice, blogging, search engines, all these things have been built on top of a very simple design, which just says keep the Internet dumb and let the intelligence run at the edges."

THE UNIVERSE AS COMPUTER

If quantum mechanics were a singer it would be James Brown.
Quantum mechanics is the James Brown of sciences.
—SETH LLOYD

What is the ultimate computer? The biggest, baddest calculating machine we could ever produce? When futurists talk computers, they inevitably focus on the computer of their dreams (drumroll, please): the quantum computer.

And they have a good model of just how powerful such a computer could be: the universe itself.

Yes, believe it or not, "the universe is computing," says Dr. Seth Lloyd, professor of quantum mechanical engineering at MIT and author of *Programming the Universe: A Quantum Computer Scientist Takes on the Cosmos.* From the very beginning of time to the present day, the universe has been creating itself in much the same way that a quantum computer works. "These little tiny quantum fluctuations that tell the universe to do this or that say, 'Let's form a galaxy here,

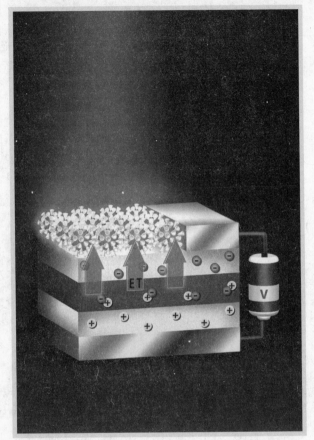

Behaving much like a fluorescent light but much more efficiently, wireless nanocrystals radiate visible light. Courtesy of Los Alamos National Laboratory.

or let's split this piece of DNA here over in this other place'—these little accidents programmed the universe.

"And it's this process of programming the universe with quantum fluctuations that gives rise to the computation we see around us, which produces all sorts of complexity and structure and beautiful things and horrible things, and most of all, amazing things."

HACKING THE UNIVERSE

Lloyd's understanding of the universe as a quantum computer stems from building tiny laboratory quantum computers for almost a decade and "coaxing individual atoms and photons to store bits and to compute and watching them work their magic. And in doing that, I realized that essentially, not only the atoms that we're trying to build quantum computers out of but [also] every single atom out there, and every photon, every electron, every elementary particle carries with it bits of information. And whenever they collide or bonk off of each other, those bits flip. So the universe is actually already computing, and it's storing information in the microscopic motions of everything—the vibrations of the air, the vibrations of radio waves—and every time those vibrations change or those bits flip, the universe is computing. That's why we can build quantum computers. We're actually hacking into the ongoing computation that the universe is performing."

Only a physicist—or should I say "ultimate geek"?—would brag about hacking the universe. Lloyd says that if we can just get a handle on how all these calculations are made by the universe, how to coax the molecules and atoms and photons of our universe, how the universe did that to make everything that we have now—we'll understand much more about computers and maybe even be able to build a giant quantum computer someday for ourselves.

"In fact, in order to understand the way in which the universe computes, we actually have to build quantum computers. If you talk to them very nicely and ask them very politely, essentially by shining radio waves on them, you can get those bits to flip and to perform computations. But you can't make them compute unless you understand very well how nature is computing already."

The idea that every atom in the universe registers bits of information dates back into the latter part of the nineteenth century. "This was a discovery made by the great statistical mechanisms [of] James Clerk Maxwell in Cambridge and Edinburgh, Ludwig Boltzmann in Vienna, and Josiah Willard Gibbs at Yale—they came up with the formulas to describe this funky quantity called entropy. Which, till then, had been known as merely something that stuck up the wheels of steam engines and caused them to do less work than you might want them to do. They realized that entropy had to do with the microscopic motions of atoms and molecules, and, phrased in modern terms, the formulas they came up with to describe it meant that entropy was the number of bits of information required to describe the motions of these atoms and molecules. And then Boltzmann went and constructed his equation, called the Boltzmann equation, which actually describes how those bits flip when molecules and atoms collide."

IT'S NOT YOUR AVERAGE PC

Lloyd's laboratory quantum computers are very crude, made out of about a dozen atoms strung together in a molecule. And they don't work at all like your PC or Mac. "Quantum computers don't run Windows yet. In fact they run something more like Linux, a very basic version of Linux."

Your PC works by breaking up the information you type in, speak, or put in with your joystick when you're playing Tetris. "It busts it up into the smallest possible components called bits. A bit is a distinction between a zero or a one, or in a computer, a switch that's open or closed, or yes or no, on or off, heads or tails. It's the smallest possible chunk of information. In a conventional computer, a bit can be stored by pouring a bunch of electrons into a bucket called a capacitor."

That bucketful of electrons would represent a 1. When you empty that bucketful, the empty bucket would represent a 0. So you've switched the bit from a 1 to a 0, the binary language of computing.

CUE THE QUBITS

A quantum computer works by moving electrons around too. But instead of a whole bucketful of them, to represent a 1 you just have a single electron. That's not so strange; it's just taking the ordinary notion of how a bit is stored in your PC and shrinking it down to the level of a single electron.

But then something very funky happens next, because quantum mechanics is very weird. It's very counterintuitive and strange: One of the central counterintuitive features is that an electron can be both here and there at the same time.

That's not a misprint. In quantum mechanics an electron *can* be both here and there at the same time. Don't ask how; no one knows. Just take it on faith, or rather on science. It's been tested and proven over and over again to be true. "If quantum mechanics were a singer, it would be James Brown. Quantum mechanics is the James Brown of sciences," says Lloyd.

"So if the electron is over here—and that's one—and the electron over there is zero, then the electron is here and there at the same time, in some funky quantum mechanical sense. That's a bit that registers zero and one at the same time, a so-called quantum bit, or qubit [pronounced *cue*-bit]."

Now how do you make use of that if you don't know which one it is? Don't we have to know where the bit or qubit is a 1 or 0 to store data?

"If only life were so certain, that would be great," Lloyd says. "In fact, sometimes not knowing what's going on is helpful. I certainly find that in my own research. And quantum computers use that in a very simple but nice way, but still counterintuitive, to do things classical computers can't.

"And a good way to think of this is to imagine what does a bit mean? A bit on its own doesn't mean anything: zero, one, yes, or no. It depends on what it does." A bit may also represent an instruction. "So a bit in a computer could tell the computer [to] add two plus two. Or tell the computer to add three plus one. And if you take a quantum computer and you feed in this funky quantum bit, this qubit that reads zero and one at the same time, then it instructs the quantum computer to do this and to do that at the same time. Say, to add two plus two and to add three plus one at the same time. That's something no classical computer could ever do."

This multitasking—the ability to do many things at the same time—is what gives quantum computing its real power. "The ability to do many things at once allows the quantum computer to explore many, many, many more possibilities than any classical computer could," says Lloyd.

BREAKING THE CODES

And just what do you do with all of this computer power?

"Code breaking is the killer app of quantum computation," Lloyd says. *Killer app* is another geeky phrase, a high-tech rendering of the ultimate computer program—that is, the killer application, the one so good that it does away with all the competition. Get it? We'll let Lloyd illustrate. "With a relatively small quantum computer with a few tens of thousands or hundred thousands of qubits able to form a few billion operations—which is peanuts for an ordinary computer—you would be able to break all the public key codes that are used, for instance, to send information securely over the Internet. And it's not surprising that the NSA [National Security Agency] and the CIA [Central Intelligence Agency] and other three-letter agencies are very interested in quantum computers."

But wait! How happy am I to hear that a small quantum computer can break all my secure passwords and steal my identity? Not very . . .

"Luckily, before you decide that you're not going to buy your coffee over the Internet anymore, quantum mechanics actually supplies a solution to this problem," because if you store bits of information as quanta, "you can actually distribute information in a way which is provably secure—or at least it's guaranteed by the laws of physics themselves. So in order to break these codes, you'd have to mess around or discover new laws of physics."

That makes me feel better. I like the laws of nature.

"So quantum computers not only create problems, at least for the

NSA—though I like to think of breaking codes as solving problems—[but] they also provide technological solutions that will allow us, the world, to continue. And in fact, quantum cryptographic systems are available today. If you are willing to write a big enough check, you can buy one for your house and feel very secure indeed."

But don't start writing that check just yet. You might not have a place to put your quantum machine.

"If you are willing to reinforce your desktop, we could put one on it right now. The ones we have over at MIT look like a giant beer keg cooled with liquid helium to cool the superconductor magnets in them. And snaking out of them are wires that connect us to about a million dollars of electronics.

"But in fact, because there are many different ways of building quantum computers, and because they're getting more powerful all the time, and the technologies for miniaturizing the components—just like the technologies for miniaturizing components of regular computers—are advancing, you might be able to get one on your desktop in a decade or two—as long as you don't mind it doing many things at once rather than one thing."

A QUOOGLE IS BETTER THAN A GOOGLE?

One tool suited to the multitasking tools of quantum computing is a search engine. "I've been fooling around with quantum versions of Google, which we would call Quoogle or something like that," Lloyd says. Imagine how fast that engine might run. Other researchers talk about using quantum computers to create better weather and storm forecasts or model global warming. That brings us full circle, because one task that is truly worthy of the supermuscular computer is understanding the universe, how it evolved and works at its basic and most fundamental scales.

"To understand what happened in the first few instants of the big bang. To understand what happens when black holes evaporate. To understand what happens when you have complex quantum system—constructed gazillions—that's another technical term—of atoms or of elementary particles. So with Dave Corey [Ph.D.] at MIT, we've constructed these quantum simulators—or you might [even] call them quantum analog computers—because they're analogs of other physical systems we want to simulate. And we're able to simulate all sorts of weird quantum effects that you could never capture on a classical computer."

Quantum computers "are essentially little laboratories that allow us to create quantum weirdness and explore its features," Lloyd says. And the quantum world is certainly full of weird effects. Take something called "spooky action at a distance." This is not science fiction. *Spooky action at a distance* is a phrase coined by Albert Einstein for an action that almost defies reality but is common in the quantum world. It occurs when two particles become "entangled" in a quantum way. If you "touch" or change the state of one of the particles, the other particle is changed instantaneously—at the exact same moment—even if the other particle is at the other side of the universe, seemingly—but not actually—violating the laws of the speed of light.

Quantum computers allow you to explore this weird but real phenomenon. "I recently, together with some other researchers, found that you could use this spooky action at a distance, this funny quantum entanglement, to escape from black holes—assuming you happen to find yourself caught in one," Lloyd says.

Good to know that, just in case . . .

"So if you have this funky entanglement between the inside and outside of a black hole—which you do because Stephen Hawking showed decades ago that black holes radiate—and, in fact, this radiation that comes out of a black hole is entangled with what's going on inside the black hole," then under the right conditions, something that falls into the black hole—a spaceship or TV set, or anything else—might be able to escape in a transformed state. "Indeed, I've had several computers that ended their lives as black holes. So you might even be able to use a black hole itself as a computer if you could figure out the right way to program that." Once again, we're back to the concept of the universe as a giant quantum computer. But be careful about setting up this experiment. "As this theory has not been tried out in practice, I feel obliged to warn not to try this at home yet. Do not jump into a black hole just now. We can't guarantee the results," Lloyd says.

BEAM ME UP, LLOYD

But it gets even weirder. This process of escape from a black hole—given that it does work—works by a process familiar to millions of fans of Captain Kirk. "A process akin to quantum teleportation—as on *Star Trek*. This is something that could easily happen on *Star Trek*; Kirk is teleported, falls into a black hole, and teleports out at the last moment, before he hits the singularity." In fact, using entanglement, Lloyd and his colleagues are "looking into the possibilities of teleporting a rubidium atom from one place to another." That's a far cry from beaming Kirk down to Vulcan, but it's a start.

"That's a nice feature of quantum computers. They're a laboratory for exploring quantum weirdness in the universe."

Lloyd also views quantum computers as "being kind of poetic. After you've talked with atoms for a while and get on their wavelength and learn to listen to what they're saying back, what they're saying has a certain strange and unearthly poetry of its own. And, indeed, you might imagine trying to use a quantum computer to compose a poem that says many things at once. Except, in fact, poetry often does that. One of the beautiful things about poetry, of course, is that words have many meanings. And so maybe we could say that poetry is already kind of quantum mechanical."

When quantum computers become part of everyday life, like your PC is now, how else might it influence the way we see ourselves?

"Every time human beings have made a new kind of technology

or machine, it has always ended up transforming human beings in ways that were unexpected and very hard to predict. As a scientist and a quantum mechanic, for me one of the most remarkable machines was the clock," Lloyd says. "When clocks were first invented, they had a huge effect on how people saw the world. And they began to see the world as if it were clockwork. And although that sounds kind of silly now—the world is clockwork—this mechanistic view of the universe, the idea that it's a machine, was extremely powerful, and indeed, you can think of it as a basis for all science. Particularly looking at biology right now, which is uncovering the mechanistic mysteries of cells.

"So I actually think that in learning to view the universe not just as a machine but [instead] as an information-processing machine, and specifically as a quantum information–processing machine, we're likely to see the world very differently and come to new understandings of it. And I don't know what those understandings are going to be.

"For me, those understandings count for a better understanding for, you know, how to build quantum computers, and how to make them compute. And then maybe also we can understand things about how life began or how the universe became so complex by looking at quantum computation."

When may quantum computing become more than a laboratory curiosity?

"I think that these giant quantum computers that are going to break these codes and strike fear in the hearts of the NSA are a decade or two away, at the earliest. It's very hard to predict technological progress. I think that it's far too early, for instance, to start some kind of mini Manhattan Project to build these. I think we'll be better off having, sharing openly our scientific advances with other scientists around the world to build them.

"However, these quantum simulators that can simulate chunks of the universe and that can construct new understanding of how complex quantum systems can behave, we can build simple versions of those already, ones with a few billion billion atoms, and investigate the properties of matter.

"And I think that for simulating stranger things, and for investigating quantum weirdness, I think we're just going to have a string of ever more powerful quantum computers. And indeed the number of quantum computers in the world has gone up by a factor of a hundred or so. People are computing with atoms, molecules, superconducting circuits, quantum dots, electrons, et cetera. Basically, anything out there that you can shine light on in the right way, you can make it compute.

"So it's a pretty exciting time for quantum computing."

PART XI

—

BEAUTY IN THE
DETAILS

CHAPTER TWENTY-EIGHT

THE JOY OF KNOWING

The scientist does not study nature because it is useful; he studies it because he delights in it, and he delights in it because it is beautiful. If nature were not beautiful, it would not be worth knowing, and if nature were not worth knowing, life would not be worth living.

—JULES HENRI POINCARÉ

There is an old saying about art: The beauty is in the details. Why can't the same be said about nature or technology? As the late physicist, practical joker, and storyteller Richard Feynman put it in a 1981 interview on the BBC program *Horizon*:

I have a friend who's an artist and he's sometimes taken a view which I don't agree with very well. He'll hold up a flower and say, "Look how beautiful it is," and I'll agree, I think. And he says "You see, I as an artist can see how beautiful this is, but you as a scientist, oh, take this all apart and it becomes a dull thing." And I think that he's kind of nutty.

First of all, the beauty that he sees is available to other people and to me, too, I believe, although I might not be quite as refined aesthetically as he is; but I can appreciate the beauty of a flower.

At the same time, I can see much more about the flower than he sees. I can imagine the cells in there, the complicated actions inside which also have a beauty. I mean it's not just beauty at this dimension of one centimeter, there is also beauty at a smaller dimension, the inner structure.

Also the processes, the fact that the colors in the flower evolved in order to attract insects to pollinate it is interesting. It means that insects can see the color. It adds a question: Does this aesthetic sense also exist in the lower forms? Why it is aesthetic? All kinds of interesting questions which shows that a science knowledge only adds to the mystery and awe of a flower. It only adds; I don't understand how it subtracts.

To be sure, I'm no Feynman. But you don't have to be a rocket scientist to appreciate his or Poincaré's vision of beauty. Being kept in the dark about the world's mysteries is not my style. I want to know what makes things tick and why.

As a journalist, I've found the joy is in uncovering the beauty in the details. Over the years, thousands of people have written me seeking the answers to the everyday mysteries that abound in science, nature, and technology. Beauty that waits to be uncovered in the workings of simple things you use or see every day, such as the unusual behavior of the shower curtain that always billows inward and sticks to my legs. Why is that? Or the bubbles in a glass of beer: Why do they appear to sink? Some of our most common experiences— things that we take for granted, such as flying in an airplane—have a simple explanation that those of us who see the beauty in nature want to know more about.

Since there are so many of them—and this book is not here to explain them all—I've assembled just a few of the experiences I've had the joy of exploring with scientists who share the joy of knowing.

THE CASE OF THE MICE CURED OF DIABETES

In the field of observation, chance favors the prepared mind.
—LOUIS PASTEUR

The history of science is full of surprises. Research that's supposed to work but doesn't. Research that starts in one direction, takes a turn, and ends up going someplace else, often yielding unexpected results. From Louis Pasteur to Alexander Fleming, from Typhoid Mary to Legionnaires' disease, science is unpredictable. A modern case in point is the work of Dr. Denise Faustman and her quest to find a cure for type 1 diabetes, the kind that destroys the pancreas of kids and teenagers and leads to lifelong daily insulin injection and blood testing.

Dr. Faustman, associate professor of medicine at Harvard Medical School and director of the Immunobiology Laboratory at Massachusetts General Hospital in Boston, announced in 2001 that she could cure type 1 diabetes in mice once she was able to stop the immune system from attacking the pancreas, which produces insulin.

"In 2001 we were doing protocols to try to reverse autoimmunity in these end-stage mice, so that we could do islet cell transplants. Remember, islet cells are the cells that secrete insulin," Faustman says.

The aim was to transplant healthy pancreatic islet cells into mice with diabetes, in the hope that the transplant would take and the new cells would help the pancreas make insulin again. There was one major problem with this approach: It's difficult to keep transplanted islet cells alive and working because the autoimmune disease, by its very nature, attacks the body's own healthy cells. In the case of diabetes, white blood cells—the T cells—attack the pancreatic cells. Dr. Faustman's aim, in 2001, was to knock out the autoimmune disease so that the transplanted cells could go to work and bring abnormal sugar levels in the mice under control.

"And that was where I got the first jarring data that the mice taught us: In these very end-stage mice, we could reverse autoimmunity. We had done the islet transplants, the animals were normal glycemic, and bingo—the big surprise was that when we took the transplant out, the blood sugar stayed flat. And that was like 'How could this happen?' and sure enough, how it happened was that islets had reappeared in the pancreas."

It appeared that once the autoimmune disease could be silenced, the pancreas could regenerate. There was no need for the transplanted cells anymore. But how did the pancreas recover? And why? Faustman suspected that adult stem cells, circulating in the blood, helped regenerate the pancreas. But she wasn't sure. Stem cells have the capability of becoming almost any cell in the body. But her discovery was so astounding that researchers doubted she had actually cured lab animals of diabetes, doubted that the pancreas had actually regenerated. She dared not speak that way in the research paper she published.

"In 2001 we weren't allowed to use the word *regeneration* in that paper. It was restoration of insulin secretion."

Faustman pressed on, searching for a reason why the pancreas could recover on its own. In 2003, she published a paper saying that perhaps the spleen was the source of those cells. According to her paper, when she injected spleen cells into her diabetic mice, those spleen cells gave rise to new islet cells, showing that not only could they "do it repeatedly in these end-stage animals but [also] there appeared to be multiple mechanisms for why these islet cells could reappear in the pancreas. And one way was without any stem cell transplant; another way was actually putting back in precursor cells from the spleen."

But in March 2006, three independent studies in the journal *Science* both succeeded—and failed—to replicate Dr. Faustman's work. The success: all the teams cured some of their diabetic mice, indicating that there was growth of new pancreatic islet cells. The failure: no evidence that the injected spleen cells were the reason or that the spleen cells gave rise to new islet cells. To Dr. Faustman and her colleagues, the specific failure was unimportant. What did matter was that for whatever reason, the animals were cured of diabetes.

"I must say when I saw the three papers that are, to a great extent, confirmatory of the major results that Denise found over the last four or five years, I was very pleased," says Dr. David Nathan, professor of medicine at Harvard and director of the Diabetes Center at Massachusetts General Hospital. "The three papers could not confirm what Denise had seen," says Nathan. "But nevertheless, there are new islets that are secreting insulin and sufficient to cure diabetes. I think that the question of mechanism—where these cells come from—remains an open question."

Faustman agrees. "Yes, so there's still debate of how this regeneration/rescue occurs."

Does Faustman still think there are stem cells floating around in the blood that help an injured pancreas repair itself?

"Yes, absolutely. I think the pancreas is still smarter than us. And I think that it can probably heal by multiple mechanisms."

She points out that research from another group in Switzerland shows that in human fat tissue, "they had found a similar precursor cell" grown in laboratory dishes that "could form human islets. So these stem cells may be in multiple locations." That means that if you stop the disease in its tracks, you may also rescue any remaining islet cells in the pancreas.

None of this speculation, says Dr. Nathan, should detract from the fact that mice, for the first time, had been cured of type 1 diabetes. "I think what we need to keep in mind is that the diabetes in these mice, this mouse model of autoimmune diabetes, was thought to be irreversible, incurable at the stage in which it was being studied."

SO MUCH FOR MICE; WHAT ABOUT PEOPLE?

Now for the $64 question: How soon can such a cure be tested in humans? The mice were given a simple and safe compound of chemicals to knock out their immune system. Could that same brew be given, safely, to people?

"It needs to be determined," cautions Nathan. "It's worth pointing out that Denise's work and the work in *Science* that's being published is all in mice. And so the mice community has reason to celebrate: Diabetes is curable in them. But we still need to demonstrate whether these lessons we've learned in the mouse are translatable to humans."

Nathan remains enthusiastic. "It is conceivable that one can use a relatively simple way of manipulating the immune system, at least as the first step, to get rid of that part of the autoimmune response. Deplete these kinds of cells that don't recognize self from nonself, and that makes, of course, for the ability to do some fairly simple early clinical trials."

Of course *early* is a relative term. "People have wondered: 'This is about three years old; how come you haven't started?'"

"Technical difficulties" remain to be ironed out first, he says. It turns out that one of the fundamental parts of Faustman's studies is that she recognized that there are these abnormal T cells—that is, white blood cells that circulate in the blood and that seem to be the cells that attack the islets. "We have to be able to measure those cells in a reliable fashion in humans before we can go on to human studies. And Denise has spent the last couple of years trying to develop these methods and translate them from measuring them in [mice] to measuring them in humans. As soon as she's developed that tool, then we can go ahead and do the first part of the clinical studies that we want to do."

Faustman says she is ready to meet the challenge. "Yes. There's a great analogy that people with diabetes will understand. If David Nathan was working in 1920 and he announced over in his research lab, down the street here, that he had discovered insulin, but he didn't know it regulated blood sugars, the chance of insulin ever working in humans and not being able to check a blood sugar and dose it is about zero. So we feel we're in the same boat, and to really make these compounds work in humans, we have to have a blood-monitoring tool. So the big job assignment that comes back to my laboratory here is taking human blood, isolating these cells, and proving we can count them and quantitate them, before David ever starts the trial over in the clinic."

Of course when you think about diabetes, an autoimmune disease, you think about the other autoimmune diseases—diseases where the body's immune system attacks the body itself, thinking the tissues are foreign invaders. Why can't the same kind of research apply to curing them? No one who works with diabetes misses that point.

"That's actually why David and I are excited about bringing this forward to the clinic, because as we were doing mouse experiments, other scientists worldwide were looking at the same bad T cells in other forms of human autoimmunity. And it turns out [that] very similar T cells that die with these same compounds have been found in lupus and scleroderma and indirectly in subsets of patients with Crohn's [disease] and rheumatoid arthritis. So there may be subsets of autoimmune patients beyond type 1 diabetes [who] might benefit from this same therapy. Obviously, in those patients you don't need to regrow their islets or rescue their islets, but you do need to rescue their target tissues that are being destroyed in different forms of autoimmunity."

Is there a risk, though, that this therapy might suppress parts of the immune system that you don't want to suppress?

"Well, that's the good news," says Faustman. Past human trials of treatments for type 1 diabetes have involved the shotgun approach of giving nonspecific immunosuppressive drugs. "And indeed, if you add nonspecific immunosuppressive drugs, you can halt the disease or slow it down, but you're also halting the good white blood cells and good T cells. The compounds we're using appear to only selectively kill the bad T cells, so the most simplistic way to view these compounds is kind of like an antibiotic. You take an antibiotic and it only kills bacteria; it doesn't kill your cells. So the antibiotic-type drugs appear to have specificity for only killing the disease-causing cells—at least in the mice and in tissue culture in human autoimmune cells."

But wait. We're getting ahead of ourselves, talking about human cures. Nathan points out that a lot of table setting needs to be done first. "Our first step in human studies, which we're not quite ready to start, will be looking at whether we can suppress these cells. We never thought, I think, when we designed the very earliest human studies, that they would necessarily cure diabetes. I think that's really far down the road. But the first step will be to see whether we could duplicate what Denise is showing in the mouse at least with regard to depleting these bad-actor cells. We have a protocol actually to do that very first step. But it is the first step in what will probably be a sequence, or maybe a long sequence, of studies before we get to the point of curing anyone with diabetes."

CHAPTER THIRTY

✳

THE MISBEHAVING SHOWER CURTAIN

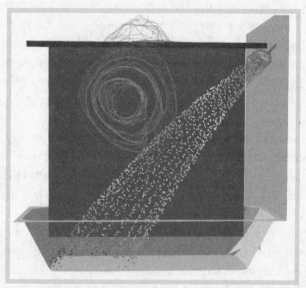

A mini hurricane in your bathtub: That's what happens when the water from the shower head creates a swirling movement of air in the tub and sucks in the shower curtain. That's why the curtain always billows in and sticks to your leg. Courtesy of David Schmidt, University of Massachusetts.

So you step into the shower and turn on the tap. But try as you might, you can't begin to scrub yourself properly all over. Your shower curtain keeps billowing in the wrong direction, *into* the tub, and

299

sticking to your body—grasping your arms, grabbing at your legs, blowing in your face. You're forced to put it outside the tub and trade a wet floor for a chance to take a shower without being groped by a piece of plastic. To you, especially if you're running late, this is a big annoyance. To physicists and engineers, it's a big unsolved problem. No one knows exactly why it happens. But the details suggest you've got a mini hurricane in your bathtub!

In 2001, David P. Schmidt, assistant professor of mechanical engineering at the University of Massachusetts Amherst, was honored with an Ig Nobel Prize for physics for publishing a paper entitled "A Partial Solution to Why Shower Curtains Billow Inwards." The Ig Nobels are awarded annually by actual Nobel laureates, one week before the real Nobel Prizes are announced. All Ig Nobel Prize winners are actual scientists who've done real science that's been reported in a published article or presented to other scientists for feedback. Like Schmidt's paper, Ig Nobel research sounds ridiculous, but the prizes are meant "to make you laugh, and make you think."

Schmidt gamely accepted his Ig Nobel wearing a shower cap and fessed up that "my own shower curtain doesn't billow inward. If I want to observe the phenomenon, I have to go to my mother-in-law's house." But Schmidt picked his paper's title because he says the problem sounds much simpler than it really is. He's part of an international community of researchers puzzling over why shower curtains have that pesky tendency to billow in and stick to your arm or leg. So far, there are mostly theories, no complete answers.

Schmidt is a founder of Convergent Thinking, LLC, a software firm that makes computer programs that mimic how sprays, such as shower sprays, act in real life. He is an expert on sprays, fluids thrust out of small openings into an unpredictable environment. Schmidt tries to predict what happens next. Most of the time, he looks at how sprays work in combustion engines, such as the diesels in trucks and airplanes and the gas turbines in automobiles that carry countless people and cargo every day. By learning more about how sprays work, Schmidt would like to find how much pollution is formed in an engine. Then he'd know which kinds of engines could help reduce global air pollution.

So far, Schmidt has found that a spray and the nozzle it bursts out of affect each other in very complicated ways, especially in a high-speed spray like the one you get blasted with if you turn your shower tap on high. Once the water bursts out of the shower nozzle, it breaks up into droplets. But what happens next, Schmidt says, is anyone's guess. The droplets can break down into even smaller mini droplets or they can collide with other drops. They can evaporate and turn into heat. They can get swirled around in eddies in the warm air in your shower.

To predict the unpredictable, Schmidt works with computer simulations of sprays that he has designed himself. As a "fun exercise" and a pleasant break from studying fuel injection, he decided to tackle the familiar shower curtain problem. He set up a mathematical model of his shower curtain on his home computer, dividing the curtain into 50,000 tiny sections. He simulated the shower running for 30 seconds. He ran yards and yards of numbers—just like the opening credits of *The Matrix*. After two weeks, he had part of the answer, something he could prove. It boiled down to this: When you step into a running shower, you step into a mini "horizontal hurricane."

The way the shower spray and the surrounding air interact, says Schmidt, "a vortex sets up, a hurricane of air turned on its side. In the center of the spinning column of air, just like in the eye of a natural hurricane, there's low pressure, and it pulls inward near the middle of the curtain." But besides the hot spray and warm air, there's the added factor of the shower curtain itself, something a real hurricane never has to consider. Schmidt says that "because of the way tension works in a hanging curtain, because the curtain is restrained by the rod, and because the bottom of the curtain is much freer to move, the bottom is pulled inward." Soon you find yourself taking your shower *with* the curtain. If you feel sorry for yourself because your shower is a daily struggle, try to save some pity for Schmidt and his colleagues, who are still trying to figure out the details of what's going on.

WHY AN AIRPLANE FLIES: DEBUNKING THE MYTH

Fly the airplane. Don't hit anything.
> —THE PILOT'S TWO RULES
> OF FLYING

Air travel these days is a harrowing experience. From the full-body frisk to the shoe X-ray to the nonexistent food (maybe that latter part is not a bad thing?), getting from here to there by air is frightening, risky, and unnerving. And that's even before you set foot on that airplane. So why make the rest of the trip even more unnerving by worrying about the actual flight? Yes, but what miracle of nature keeps us aloft?

It's got to be something to do with the air, right? There must be some lifting force to make the airplane rise off the ground. There is, but it's not very obvious. And because you really can't see the air at work, aeronautical experts have had a tough time explaining to the public why an airplane flies. Right from the time that air travel came to the masses—around World War II—teachers were looking for a quick and simple, easy to remember, handy-dandy, user-friendly, not-too-complicated way to explain why airplanes fly to a public that was about to trust life and limb to some heavier-than-air contraption. So somebody—I don't believe anyone knows who—came up with the explanation that still echoes through classrooms today.

A MYTH IS BORN

The explanation goes like this: Look at the typical textbook diagram of the cross section of a wing below. (Diagrams like this one have appeared in countless explanations of why airplanes fly.)

What do you notice about the wing? The top side is curved while the bottom is straight. This shape is very important to this classical explanation. Since the shortest distance between two points is a straight line, the distance from the front of the wing to the back is shorter on the bottom side of the wing than the top. If you were to follow the paths of the air flowing over the wing, you would notice that air particles hitting the front of the wing at the same time will split, some on top of the wing and some on the bottom. Since the top of the wing is curved and therefore its path is longer, in order for the air on top to reach the back of the wing at the same time as air particles on the bottom, they have to move much faster.

This means, in effect, that the air on top is being stretched thinner to make the same amount go farther. In physics, Bernoulli's principle says that air stretched thinly on top exerts less pressure

Classic—but wrong—illustration of airflow around wing during flight: Air flowing this way would provide no lift. Courtesy of David F. Anderson and Scott Eberhardt, Understanding Flight, *McGraw-Hill, © 2001.*

than air packed thickly under the wing. So you'd have greater air pressure on the bottom of the wing, giving the plane "lift." You might look at it in a different way and say that the thin air on top is "sucking up" the wing in the same way in which you suck up soda in a straw.

As I said, that's the classic textbook explanation that has been taught for decades. And it's quite wrong. Why? Let's take it point by point.

"One common myth," write David Anderson and Scott Eberhardt in *Understanding Flight*, "is the principle of 'equal transit times.'" That's the myth that says the particles of air—top and bottom—must reach the back of the wings at the same time. "But in reality, equal transit times hold only for a wing without lift." And we all know no plane is going anywhere without lift. So much for that myth.

Now for that idea that the shape of the wing is important. Dave Anderson points out that high-performance acrobatic planes and helicopters have wings whose top and bottom sides are identical. No rounded top and flat bottom. In fact, says Anderson, your wing could be the shape of barn door or a garbage lid. It's not the shape of the wing that gives it lift but rather how smooth it is and in what direction it's pointed. That's because a wing develops lift when air is smoothly flowing over the top of it and is forced downward toward the ground by the tilt of the wing. It's the ability of the wing to make the air flow over the top and down to the ground—the angle of attack—that

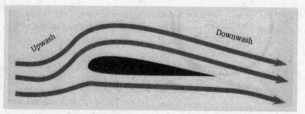

Correct illustration of wing with lift shows the downwash—that is, air headed toward the ground. Air goes down, plane goes up: Newton's third law of motion. Courtesy of David F. Anderson and Scott Eberhardt, Understanding Flight, *McGraw-Hill, © 2001.*

results in an upward force on the wing. It's Newton's third law of motion: Every action has an equal and opposite reaction. Air goes down, wing goes up. And Newton's second law of motion tells us that the greater the amount of air being diverted downward, the greater the lifting force. Sounds simple!

And you can stand under a helicopter and feel the wind blowing or watch those videos of helicopters landing in the desert or rescuing folks from boats. There the wind is quite visible.

As for Bernoulli's principle, it does play some role in shaping the way the air flows over the wing. But Anderson says that the "suction" of Bernoulli is so weak that a tiny single-engine plane such as the Cessna 172 would need to fly 400 miles per hour just to get off the ground if it relied on Bernoulli's principle for lift. My flight instructor would have a fit if I tried to fly a Cessna Skyhawk that fast; its normal takeoff speed is around 55 miles per hour, fully loaded. Another myth busted.

THE COANDĂ EFFECT

There is a fascinating detail about how the air is diverted downward on the wings that elegantly explains why airplanes fly. For a plane to fly, the wings have to attack the air at an angle that forces the air downward. One would assume from looking at how the wing is angled that the air should be deflected downward the same way that tennis balls bounce off a racket: bouncing off the bottom of the wing. But that is not the case. The air is diverted downward by something called the Coandă effect, something we experience every day but don't even know it exists.

Henri-Marie Coandă was a distinguished aeronautical pioneer who began experimenting and designing airplanes in 1905. Born in Bucharest, Coandă is probably best known for his *Coandă-1910*, an "air reactive" airplane that was powered by what we would today call a jet engine.

Coandă noticed that in testing his flame-spewing jet that the smoky gases that came out of the back of the engine tended to stick to the sides of the plane. This was a puzzlement. Why did the gases hug the fuselage? He investigated and noticed that any stream of gas, be it burning like that in the exhaust, or just your ordinary flow of air, tends to follow the curvature of a smooth surface as long as the surface doesn't have to make a sharp turn.

You've seen this happen many times. When you pour milk out of a glass into your bowl of cereal or cup of coffee, if you tilt the glass of milk but don't pour it very quickly, it dribbles down the side of the glass or cup and makes a mess on the table. Or notice what happens with your wet hands as you hold them while you're looking for a towel. What happens to the water? The water flows down your arms. But it doesn't drip to the floor! *It follows your elbow around to your upper arm and then drips.* It follows the smooth curve of your elbow. That's the Coandă effect in everyday life. Nice little detail, huh?

The same thing is happening on the wing of an airplane. The air is hugging the curvature of the top of the wing. And because the smooth wing tilts back and down, the air follows that path too. It is forced down to the ground. In fact, when pilots take off, they have to rotate the nose of the plane upward to increase this angle of attack, directing even more air downward and increasing lift.

The plane also achieves lift because the leading edge of the plane is higher than the back (trailing) edge. No doubt when you were young and eager to experiment, you noticed this effect when riding in a fast-moving car. When you hold your hand horizontally out the window with your fingertips pointed slightly up, your hand takes off like a plane.

Notice: it is the air on the top of the wing that is doing the useful work. Not the air on the bottom. The air flowing beneath the wing is not bouncing off the wing bottom, like a ricocheting marble, and providing lift that way. In fact, the airflow below the wing is so unimportant to lift that the bottom of the wing is the place where airplane

A Learjet above the fog: The indentations in the clouds are the result of the air being forced down by the jet—the downwash—creating an equal and opposite force that provides "lift" for the plane. Courtesy of Paul Bowen.

weapons designers stick all the bulky military hardware—rockets, fuel tanks, and so forth. If that air was needed for flight, you couldn't put that airflow-disturbing hardware there.

If you look out the window of your airplane during takeoff and landing, you'll notice that the wing changes shape. It appears to grow. Flaps extend backward and beneath the wings. During takeoff and landing, they give the plane added lift because they extend the size of the wing. On really big planes, you'll notice that the flaps actually separate from the wings and extend slightly below them. The air is funneled over the top of the flaps to create extra lift. Remember, the air on the bottom of the wing is not really doing any work, so it has not lost any energy and can be used by the flaps for lift.

So why not keep the flaps deployed all the time? Too much wind resistance, called drag. Whenever lift is created, so is drag. The flaps are not needed at high speeds; they would slow the plane down and could actually be ripped off the wings at the 600-mile-per-hour speed that a jet airliner flies. (My flying instructor, Richard Orentzle, had to check on the condition of the flaps on my Skyhawk 172 when I went into a dive during stall practice with flaps fully extended. That was a no-no. It could have buckled the flaps, but there was enough margin of error built into that sturdy little plane to absorb the abuse.)

Myths die hard. And this one is no exception. Teachers continue to use Bernoulli's principle in classrooms; I got into quite an argument with my daughter's physics teacher over this. But progress is advancing slowly as enlightened textbooks and encyclopedias begin to modernize their thinking. And the Internet, that great online debating arena, is host to quite a number of discussions about the truth about why airplanes fly.

CHAPTER THIRTY-TWO

THE GREAT CHAMPAGNE BUBBLE MYSTERY

Tiny bubbles in the wine,
Make me happy, make me feel fine.
— BY LEON POBER, AS SUNG BY
DON HO

You've had a wonderful wedding, bar mitzvah, graduation, or New Year's Eve party, and now you've got a few open and partially filled bottles of champagne left over. Who would want to waste that bubbly? Not me! But is there a way to keep the brut from going bad overnight, to keep the champagne bubbly so that you might enjoy that mimosa the next morning?

This is a question that has been debated for centuries, since the French started making champagne. And the French thought they had the answer: the old spoon trick. Hang a silver spoon, handle down, in the neck of the bottle, pop the bottle into the fridge, and the booze should hold its bubbles for a few more days.

A team of Stanford University researchers put the idea to the test—all in a thirst for knowledge and digging into their own pockets for research funds. They found that the spoon theory fell flat. Would any other technique work? They decided to find out.

The idea for their test came in 1991, when a reporter from Germany called Stanford University chemistry professor Richard Zare to find out whether and how the spoon theory works. Zare, a "bubble-ologist," had just published an article in *Physics Today* about the physics of the bubbles rising in a glass of beer. The question intrigued the physicist but left him doubtful about the magical powers of a spoon.

"I thought it might be a bubblemeise," said Zare. That's a takeoff on *bubbemeise*, Yiddish for "grandmother's tale."

Then in 1994, renown food expert and author Harold McGee made a guest appearance on my radio program, *Science Friday*, where I asked him the same question. McGee, no stranger to science sleuthing, was a perfect choice for this question. He had published a paper in the journal *Nature* on why the froth of a soufflé stabilizes when you beat the eggs in a copper bowl. In McGee's words, the two friends realized an obligation to human knowledge: Here was an experiment that cried out to be done.

Rising to the challenge, they convened an informal team of eight amateur taste-testers, including Zare's wife, Susan; McGee's wife, Sharon Long, a professor of biology; two more biology professors; a law professor; and a physician. A true cross section of scientific brainpower.

In the true spirit of scientific investigation, they tested and rated 10 bottles of champagne, carefully controlled for temperature and with a single glass of champagne removed to make sure all were the same at the start. The bottles received five different treatments:

- One bottle was opened just before the *test*.
- One bottle was opened 26 hours earlier and left *uncorked*.
- One was opened for 26 hours with a silver spoon in the *neck*.

- One was opened for 26 hours with a *stainless-steel* spoon in the *neck.*
- One was opened and recorked overnight.

The results were highly unexpected.

"What we found was a surprise—at least to us," Zare said.

The spoons, silver or stainless, were not especially successful in maintaining the sparkle of the wine. But spoons and all other treatments worked better than recorking the bottles. Hard to believe, but at least in this test, recorking the bottles seemed to be the best way to make champagne *lose* effervescence and taste.

The hands-down winner? Simply leaving the bottle open—uncorked—in the refrigerator. In fact, the two bottles left open in the refrigerator for 26 hours averaged a higher score than any other treatment—including just-opened champagne!

Why such a surprising result? No one knows for sure, but there may have been a complicating factor or two. Perhaps the testing method itself influenced the results—that is, the state of the observers by the time a glass of champagne had been sipped (in some cases, more than sipped) from each of 10 bottles. As research scientists, several members of the team noticed what Zare called "fatigue of the instrumentation." In this case, the instruments (the human beings) might have—how should we put this?—been influenced by the alcoholic content after so many glasses of wine.

"Our palates were not as fine as at the beginning. Eventually we didn't feel quite right about letters and numbers," McGee recalled. "You hear of the observer influencing the observed, but not often the observed influencing the observer," McGee said. "I think we have a reverse Heisenberg principle here."

One team member had a philosophical disagreement with a test that used bubbles as the mark of quality. "I am unable to disaggregate the gestalt of the wine," he declared, setting down his scorecard.

Zare and McGee concede that their results are very preliminary and that their data set is small. As Zare puts it, "We are struggling to achieve statistical significance."

As reported by Stanford University, "After their study was completed, McGee learned of a French study that seems to confirm their results, conducted under the auspices of the Centre Interprofessionnel des Vins de Champagne. French science journalist Hervé This-Benckhard told McGee by e-mail: "I think we can affirm now that a spoon, made of silver or stainless steel or of aluminum, has no effect on what the French term *éventage*, or the loss of gas."

But the mystery of the uncorked bottle remains. Zare echoed the age-old complaint of scientists: "As usual, more research is needed, and the observations we have made open more questions about the laws of 'fizzics' than they settle. . . . We hear that it makes a difference if you do the experiment on Dom Pérignon, and we'd love to test that out."

"Our thirst for knowledge is still not satisfied," he said.

This is certainly an experiment you *can* try at home.

LIPSTICK AND CHAMPAGNE DON'T MIX

While we're on the topic, Zare also found that sipping champagne while wearing lipstick kills the bubbles. Refilling a glass with a lipstick stain on it will eventually cause the newly poured champagne to go flat. Lipstick contains surfactants that reduce the surface tension of the bubbles and cause them to pop. So the foamy fizz will eventually fizzle in the glass as the champagne comes in contact with the lipstick. Now if that detail isn't a surefire party icebreaker, the drink's on me.

AND FINALLY: FALLING BEER BUBBLES

Here's a cute bar trick. Bet someone a glass of Guinness that bubbles sometimes go down instead of up when beer is poured into a glass. If you're a beer drinker, especially Guinness, you've always suspected it. Sometimes it looks like the beer bubbles *are* going down instead of up, but how could that be? Being buoyant, bubbles always rise to the surface, don't they?

"Bubbles are lighter than beer, so they're supposed to rise upward," says Zare. "But countless drinkers have claimed that the bubbles actually go down the side of the glass. Could they be right, or would that defy the laws of physics?"

The quandary came to a head in 1999 in the "battle of the bubbles" after Australian researchers predicted bubbles would behave in this crazy way on the basis of computer models showing that bubbles could theoretically flow downward in a glass of Guinness. (Sometimes I think these guys have too much time on their hands.) Zare and former Stanford postdoctoral fellow Andrew J. Alexander were skeptical. Rising to the challenge, they decided to personally investigate several liters of the brew up close and personal.

"Indeed, Andy and I first disbelieved this and wondered if the people had had maybe too much Guinness to drink," Zare recalled. "We tried our own experiments, which were fun but inconclusive. So Andy got hold of a camera that takes seven hundred fifty frames a second and recorded some rather gorgeous video clips of what was happening."

WHAT GOES DOWN MUST COME UP

The video didn't lie: The beer bubbles did sink to the bottom of the glass. But why?

"The answer turns out to be really very simple," Zare explained. "It's based on the idea of what goes up has to come down. In this case, the bubbles go up more easily in the center of the beer glass than on the sides because of drag from the walls. As they go up, they raise the beer, and the beer has to spill back, and it does. It runs down the sides of the glass, carrying the bubbles—particularly little bubbles—with it, downward. After a while it stops, but it's really quite dramatic and it's easy to demonstrate."

And if Guinness is not your brew, the bubbles will do their thing

Circulation patterns of bubbles in a glass of Guinness Draught, confirming that some bubbles in the glass go down and not up. Photo courtesy of Richard Zare.

in other carbonated beverages too. "The bubbles are small enough to be pushed down by the liquid," said Alexander, now a professor at the University of Edinburgh in Scotland. "We've shown [that] you can do this with any liquid, really—water with a fizzing tablet in it, for example." Message to Alka-Seltzer: Take notice.

Okay, so much for winning a bet. But who cares?

Zare, an expert on carbonation, says that the bubbles of CO_2 in seltzer, champagne, and beer all float to the top differently. "It's just paying attention to the world around you and trying to figure out why things happen the way they do," Alexander added. "In that case, anyone that goes into a pub and orders a pint of Guinness is a scientist."

CHAPTER THIRTY-THREE

OPEN-SOURCE BIOLOGY

The power of the open-source concept isn't so much the cost, it's the community.

—RICHARD JEFFERSON

While to many people the computer world is divided into two camps—Mac people and PC people—there is a third camp that gets less attention: the open-source folks. These people are united to create software tools and even computer systems, such as UNIX and Linux, that are free to be used by anyone. You can download them on the Internet and run them on your computer at no charge.

That idea is now beginning to spread through the world of biology. Why not create open-source tools for biologists who can't afford to buy the name-brand tools to help in their work in genetic engineering? To help developing countries make the products they need but can't afford to buy?

"If we're going to solve any of the problems that are really aching for a lot of the world, that part of the world has to solve the problems themselves," says Dr. Richard Jefferson, chairman, chief executive officer, and chief scientist of CAMBIA, a nonprofit biotech research organization based in Canberra, Australia. Jefferson's BIOS initiative has taken the unheard of step of taking biotechnology into the open-source community.

For example, topping the list of important techniques in genetic engineering are a set of tools, one used to insert genes into a genome and another to help figure out in which cells those inserted genes ended up. These tools help you slice and dice the genome and put in new genetic information to create a unique product. But most of those tools are controlled by a few big biotech companies. And while they might be willing to license those tools to university researchers, for instance, they're not so open to handing those tools over for free to people actually planning to make seeds for farmers in developing countries where they can't afford to buy new seeds each growing season.

Jefferson and CAMBIA have been able to develop alternative tools for doing some of these key genetic engineering processes, methods that they say don't infringe on any of those patents owned by the big guys. And they're offering to make those tools freely available, free to others to use or expand, sort of the biological equivalent of open-source software, like Linux and the others. You can use them for free. You can improve them, but you have to put them back into the public domain so other people can use them, because if you wait for others to give you the tools, you may have to wait for a long time.

"When you talked about open source being basically tools for free, the real issue is that they can be designed and built and tuned by a large community. So the power of the open-source concept isn't so much the cost, it's the community."

One project Jefferson has been working on for decades involves giving the worldwide community the open-source tools and methods it needs for transferring genes into plants. It improves on the standard, patented method of using a common bacterium to carry the new genes into the plant.

"You take a soil bacterium called *Agrobacterium* that naturally lives with plants, which forms, under normal conditions, a pathogenic relationship. Basically it makes galls, like tumors, on plants. That's a normal process, been going on for millions of years. It was discovered that in that process, the bacteria are capable of transferring part of its own genome into a plant and reengineering the plant, in fact, to feed it and clothe it, as it were."

Jefferson and his colleagues wanted to make use of this talented bacteria. They wanted to genetically engineer the bacterium, to improve it. "Basically we think we could do something that works better. The downside: *Agrobacterium* is naturally a pathogen. That means that it causes diseases in plants." So since the 1980s, Jefferson and others in his field have been working to "disarm" the plant so that it no longer made tumors but still could become infected.

"We decided to try to convince or coax some very benign bacteria that naturally live in a symbiotic relationship with plants and ask them, basically by minor adjustments, if they can transfer genes into plants. And the surprising and exciting observation is that yes, they can."

But that's when he realized that a morass of patents stood in his way of introducing this method into the open-source community. "We realized that the patents claim all aspects of methodology, the materials, the tricks, the tunings, and it became what we call a patent thicket, with literally hundreds of patents. So it's not so much that it's dominated by one multinational; it's that there are so many patents involved that very often it only takes one out of this giant Tower of Babel of patents to be denied to stop the whole thing from working."

The open-source community bypassed the problem, freeing up the tools for improving the bacteria and providing a set of open-source tools that people could use to avoid the patent mess. "It turns out that this has been a need that's been very sorely felt by a lot of people in the science community and especially in the development community. There are countless excellent scientists totally committed in the developing world or, in fact, in small and medium enterprises in the U.S. and the U.K. and Europe, Australia, that are dying to get out there and start innovating on behalf of smaller markets instead of just the big-margin innovations."

And what kind of innovations might we see, now that these tools have been unleashed? Jefferson says that "part of the beauty of open source is its lack of prescriptiveness from one guy like me or one person." In other words, the sky's the limit. Take what you need.

"If people have needs that are legitimate and they think that they're not being served by existing technology, the power of open source is the ability to craft the technology depending on your needs and your view that you are not being served by existing technology."

Making the tools freely available to private individuals does not mean that such folks aren't free to profit from their innovative use of the tools. Jefferson says there needs to be a distinction between the "tools of innovation, which we feel are an absolute fundamental human right, and the fruits of innovation," which can be proprietary. For example, let's say that some young scientist makes use of these tools and she uses them to develop an improved rice strain "that she thinks has the cat's meow properties and she wants to market it with her own trademark and her own plant breeder's varieties.

"I don't see that that in any way suppresses the ability of other people to develop other rice varieties. And that's the real key. As long as it doesn't suppress free and open both competition and innovation, we see no problem with that. We don't even see a problem with Monsanto or Syngenta using our technologies and producing new strains of corn or soybean. The real issue is not suppressing other people from providing alternatives and not suppressing other people from developing small, medium enterprise that can be so exciting to us."

However, what is to prevent a large company from coming in and cherry-picking the best new ideas from this "creative commons" and taking out a patent on them, in effect going against the very people who developed them?

"The basic issue is that it's not a creative commons issue. It's a protected commons. And that protection is built into our thinking. In other words, we are not anti-intellectual property or anti-patents. We're very much for using them wisely and much more discreetly. But in a sense, a patent license is the very stick that goes with the carrot of the technology that says, 'Share nicely or you don't get to be part of the community.' And the patent licensing and the opportunity to sue for infringement has to always be maintained as part of our structure."

But wait. Sure, the small-business person can sue a large company for patent infringement. But what chance does a small entrepreneur have against a team of high-priced lawyers? Once again, says Jefferson, the open-source software community has been down this road and has an interesting answer.

"Look at the open-source software community, or what started out as the free-software community. The license that guides the development of Linux is called the GPL, or GNU Public License. And you know what? The total amount of money made on Linux is in the billions and billions of dollars, and do you know how many times that license has been litigated? Zero. Because they have an extraordinarily bright counsel named Eben Moglen at Columbia University in New York who, whenever there's threat of litigation, just picks up his briefcase, goes over and talks to these people, and gets them to understand how it's in their best interest not to. That's the beauty of this. It's not about confrontation. It's about awakening people to what's in their own self-interest and getting communities going. And it's worked for Linux. It's worked for the literally tens of thousands of other software programs, including the great Apache Web server, which drives most of the World Wide Web, which was developed by our colleague, Brian Behlendorf. These are innovations that are of staggering importance to the economy, and yet, by and large, the license guiding them has not been litigated. We think that we can do the same thing. You don't have to be confrontatory to be successful."

The same is true for the open sharing of information. An important function of the open-source community is sharing what you know—that is, once an improvement is made, that advance is placed back into public domain, where other people can use and improve on it further. But common sense says that in the business world, knowledge is power, that sharing what you know is dangerous to your business. Not so, says Jefferson. "Yochai Benkler at Yale has written a marvelous article called 'Sharing Nicely,' in which he looks at the economics of innovation, and it seems to be that we actually get faster and better innovation in an industry by sharing than we do by competing."

The end product is the capability of people to tune in to their own needs. "So instead of us talking about science done for the Third World, we should rethink [it] as science done by the Third World. Every time you see a picture of someone in Africa, India, wherever else, you see a picture of a hungry person or a poor person. But what you rarely think about is that they're a creative person, and they are. And so the issue is we have this massive untapped source of innovation, which is normal human beings that want to solve problems. And our job is not to fix the Third World. Our job is to remove the constraints to their own creativity, and that's a huge task. It means policy, economics, not just molecular biology."

PART XII

—

THE QUEST FOR
IMMORTALITY

CHAPTER THIRTY-FOUR

STEM CELLS, CLONING, AND THE QUEST FOR IMMORTALITY

From my perspective as a physician, the need for this work is now greater than ever. Stem cell research can make a difference in people's lives, and our efforts can serve as a paradigm for how we might ultimately be able to use many new developments in biology.

—DAVID T. SCADDEN, MD,
HARVARD UNIVERSITY

In the movie comedy classic *Monty Python and the Holy Grail*, a cart rolls through the streets of a plague-stricken medieval village, while the cart driver calls out, "Bring out yer dead! Bring out yer dead!" Bodies of plague victims are piled high on the cart. But one body keeps sitting up and announcing, "Not dead yet, y'know!" If you go to *Spamalot*, the Broadway musical based on the Python movie, you can buy a T-shirt on sale in the lobby that says, "I'm Not Dead Yet. . . ."

If I were you, I'd snap one up. In the future, mortality could be just another disorder that can be cured. There are two ways that could happen. You could clone your pet or perhaps your loved one or yourself. Or you could replace yourself, organ by organ, with stem cells that grow new body parts.

CLONING TO KEEP YOUR CAT FOREVER

In late 2004, a Texas woman paid $50,000 to a California company named Genetic Savings & Clone to have her late beloved cat, Nickey, cloned. Sometime later, she brought home "Little Nickey," a nearly identical, nine-week-old kitten that she swore was her old kitty reincarnated. It was the first time an American pet had been cloned. The very first animal to be cloned was a tadpole, in the 1970s. The first cloned mammal was Dolly, a sheep born in Scotland in 1996. Since then other animals have been successfully cloned: lab animals, including cats, mice, and rhesus monkeys; livestock, including mules, pigs, and calves; and even endangered species, such as a guar, large wild oxen from India and southeast Asia, and a mouflon, a wild sheep. Genetic Savings & Clone was forced to close up shop at the end of 2006, saying in a letter to customers it was "unable to develop the technology to the point that cloning pets is commercially viable." There were just not enough customers willing to pay the cost of cloning Snuffy, even when the price was dropped to $32,000.

In some businesses, though, that price may be a bargain. Thoroughbred racehorses, with proven racing ability, are now being cloned with the hope that the clones will give their owners a run for the money. A champion horse, with winnings of over $380,000, was cloned by a Texas breeder for the bargain price of $150,000.

Cloning your aging dog still presents more of a challenge. In August 2005, a team of South Korean scientists unveiled Snuppy, an Afghan hound who is the world's first cloned dog. But because female dogs ovulate only once or twice a year, and at unpredictable intervals, one Snuppy required the use of more than 1,000 embryos, plus 123 surrogate mother dogs. Only 3 became pregnant, and only 1 gave birth to a healthy puppy, Snuppy. That long string of failures means a very high price tag, until dog cloning becomes a lot more efficient.

But there are other reasons to clone besides replacing a pet or breeding a winner. The usual reason for cloning plants or animals is to mass-produce rare or desirable organisms. Cloning means making an exact copy of biological material. Farmers and vegetable gardeners know that many plants and vegetables, including grass, potatoes, and onions, clone themselves. These plants send out a kind of modified stem called a runner. Wherever the runner takes root, a new plant grows, doubling your crop.

If you have potted plants that you cultivate indoors, you've probably done some cloning yourself. You're certainly familiar with taking a leaf cutting from one plant and growing a new one. If you've ever done that, you've just cloned your original plant. (I have a Christmas cactus that is now the grandchild of the original.) You can do this because the end of your cutting forms a mass of nonspecialized cells called a callus. With the right amounts of soil, light, water, and nutrients, the callus will divide and begin to form specialized root and stem cells—growing into a new plant.

Then there's tissue culture propagation, which botanical scientists and orchid fanciers use to grow rare plants. You can take pieces of specialized roots from a plant, break those roots up into root cells, and grow them in special culture. In nutrient-rich culture, the specialized cells become unspecialized, or dedifferentiated, into calluses. With the help of plant hormones, the calluses can grow into new plants that are identical to the original plant from which you took root pieces.

Besides plants, some animals—including humans—clone naturally. Identical twins are clones. Under certain environmental conditions, the unfertilized eggs of some animals—small invertebrates, worms, some species of fish, lizards, and frogs—can develop into full-grown adults. This process is called parthenogenesis, and the offspring are clones of the females that laid the eggs.

DIFFERENT TYPES OF CLONING

Although *cloning* is often used as a blanket term, there are three separate and very different types of cloning. The first is recombinant DNA technology, or DNA cloning, which is used all the time in molecular biology labs. To make more of the same DNA fragment for study, a scientist transfers it from one organism into a self-replicating genetic element, such as a bacterium, yeast, or virus. The DNA then can be replicated in a foreign host cell.

DNA cloning would be very useful in gene therapy. The 1992 movie *Lorenzo's Oil*, based on a real case of a boy born with a rare genetic illness that wasn't discovered until the 1980s, is an excellent primer on molecular biology and the promise it offers to cure inherited illnesses. Gene therapy sounds simple: You could remove defective genes from a sick person, clone healthy genes, and use a harmless virus to convey them to the patient's cells, where they would replicate themselves and take the places of the defective genes.

Unfortunately, right now, gene therapy is far from being therapy. In 1999, 18-year-old Jesse Gelsinger, who suffered from a rare metabolic genetic disorder, died in the course of a gene-therapy experiment for which he'd volunteered. Since then, gene therapy has been going very slowly. The main problem for researchers has been finding the right gene carrier, or vector, that will convey healthy genes into a patient's cells, and allow them to begin replicating themselves. Viruses are often suitable, but of course many viruses are harmful. In Gelsinger's case, scientists used the rhinovirus, which causes the common cold. It's also big enough to carry genes. Since Gelsinger died, scientists have been working to come up with a viable substitute.

Then there's reproductive cloning, the approach used to create Dolly and other animals with the same nuclear material as an existing animal—or a deceased pet. Scientists transfer genetic material from the nucleus of a donor animal's adult cell—obtained from the samples you sent in from your pet cat—to an egg whose nucleus, along with it its genetic material—has been removed. To stimulate cell division and growth, the reconstructed egg has to be treated with chemicals or electric current. Once the cloned embryo has developed far enough, it is transferred to the uterus of a female host, where it continues to grow and eventually is born as your new kitten.

But that means your kitten—and Dolly and other animals created with reproductive cloning—aren't completely identical copies of the original donor animal. Only the clone's nuclear DNA is the same as the donor's. Some of the clone's genetic material comes from the mitochondria, powerful cells in the egg. Mitochondria contain their own short segments of DNA, and acquired mutations in mitochondrial DNA may play an important part in aging.

So cloned animals often have unexpected kinks, including much poorer health than their donors. They tend to have compromised immune systems and can be prone to tumors, infections, and other complications. Many cloned animals simply haven't lived long enough to tell us how clones age. Finn Dorset sheep like Dolly normally live to the age of 11 or 12. Poor Dolly had to be put down by lethal injection in 2003, when she was only 6. She had had lung disease and arthritis, which was crippling her. People with cloned pets often wonder why they lack some of the qualities that made the originals so endearing. Texas A&M researchers once had a sweet-natured bull named Chance. They liked him so much that they cloned him and dubbed the clone Second Chance. Unfortunately, Second Chance turned out to have a dangerous temper; he was much more of a raging bull.

Reproductive cloning also is hard to do. More than 90 percent of cloning attempts fail to produce healthy offspring, and more than 100 nuclear transfer procedures may be needed to make one successful clone. Dolly, for example, was a single success out of 276 attempts. Scientists hope to use this method to reproduce animals that have been genetically engineered to produce useful drugs, or to serve as models for studying human disease. Researchers also would like to repopulate endangered animal species or those that are difficult to breed, such as pandas. But that remains a major challenge. (Cloning extinct animals would be even more difficult, because the egg and the surrogate mother would have to belong to different species from the clone.) So far, certain animal species, such as chickens, have resisted attempts to clone them. Cloning has been limited to only a few animal species, and some species may turn out to be more resistant to reproductive cloning than others.

THERAPEUTIC CLONING

The third type of cloning is called therapeutic cloning, and it involves the use of cloning technology in medical research. One day, it may allow the production of whole organs from single cells and make waiting lists for donations of organs such as the liver, kidney, or heart a thing of the past. Or therapeutic cloning could produce healthy cells that could replace damaged cells in degenerative diseases such as diabetes, Alzheimer's, and Parkinson's.

In 2002, one biotech company reported that its researchers had successfully transplanted kidneylike organs into cows. The team created cloned cow embryos by removing the DNA from donor cow egg cells, and then injecting the DNA from skin cells of a donor cow. The scientists allowed these cow embryos to develop into fetuses. Then they harvested fetal tissue from the clones and transplanted it into the donor cow. After three months, the team reported that they had not observed any organ rejection in the donor cow.

XENOTRANSPLANTATION

Another way that scientists are trying to make therapeutic cloning work is to create genetically modified pigs and harvest from them organs for transplantation to humans, or xenotransplantation. Pigs are good candidates for therapeutic cloning and xenotransplantation because their tissues and organs are quite similar to humans'. To create what British scientists call knock-out pigs, they have to inactivate the pig genes that trigger the human immune system to reject a transplanted pig organ. They knock those genes out of individual cells, which they then use to create clones from which they can harvest needed organs. In 2002, a British biotech firm reported that it had produced the first "double knock-out pigs," genetically engineered to lack both copies of a gene involved in transplant rejection.

One potential hazard of using organs from animals such as pigs is the possibility that some viruses that live in the pigs might find their way into humans and kill the recipient. This possibility is making xenotransplants still very controversial and in need of further research.

EMBRYONIC STEM CELLS

Therapeutic cloning often is described as stem cell research because both involve growing stem cells. What scientists need to grow human replacement parts are human stem cells, "blank" cells that have not yet decided what they will become. They might become heart, or

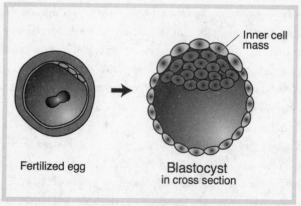

Inner cell mass

Fertilized egg

Blastocyst
in cross section

A tiny ball of cells—a blastocyst—from which embryonic stem cells are derived, a few days after in vitro fertilization: Stem cells are clustered in the inner cell mass, which is so small as to require a microscope to see.

lung, or brain cells; they are just awaiting genetic instructions. The easiest place to find and extract such stem cells in large quantities is the blastocyst, a ball of cells that forms in the first few days after a sperm and an egg join. These cells—about 32 to 200—are the foundation cells for every organ, tissue, and cell in a human body. These stem cells have not yet differentiated themselves into the more than 200 kinds of specialized cells that can become blood, neurons, skin, or organ tissue in a human body. In 1998, biologists at the University of Wisconsin reported that they had succeeded in using cloning technology—removing cells from human embryos—to establish the world's first embryonic stem line, or population, in the lab. The Wisconsin stem cells—along with most embryonic stem cells used for research—were extracted from embryos created by in vitro fertilization and were left at fertility clinics, frozen in liquid nitrogen, as spares by couples who already had succeeded in conceiving healthy children. (A small percentage of couples donates the spares to research.)

A blastocyst is no bigger than the period at the end of this sentence. Its inner cell mass contains 8 to 40 stem cells. Once the stem cells have been transferred into culture, they begin to proliferate. If after several months the original cells have grown into millions of healthy cells, without starting to differentiate into specialized cells, then they are known as an embryonic stem cell line, or colony. A line can replicate indefinitely in a lab—although stem cells do deteriorate as they age and accumulate genetic mutations. So researchers are always looking for new supplies of stem cell lines.

Adults have a small number of stem cells in many of their organs and tissues, including their blood, heart, skin, bone marrow, and skin. Adult stem cells also can be found in liposuctioned fat, pulp under baby teeth, amniotic fluid, placentas, or a newborn's umbilical cord. But adult stem cells are scarcer in the body and harder to grow in the lab than embryonic stem cells, and they don't seem to be as versatile: They might be limited to becoming cell types within the tissue where they are found. (Cord cells produce blood cells and may prove capable of generating bone and cartilage cells.)

Stem cell research for therapeutic cloning still is at a very early stage. Scientists are working to figure out how to direct stem cells to treat heart disease, leukemia and other cancers, burns, Parkinson's disease, rheumatoid arthritis, and diabetes and to regenerate nerve cells in patients with spinal-cord injuries. (Christopher Reeve, the actor paralyzed in a 1995 riding accident, campaigned tirelessly for stem cell research until his death in 2004. Michael J. Fox, the actor who has Parkinson's disease, has picked up where Reeve left off.) So far, only adult stem cells have been tested in humans, although research continues in search of treatments for many different diseases.

To generate organs or tissues for transplant, scientists would extract DNA from the patient in need and insert it into a donor egg from which the nucleus and genetic donor material have been removed. After the egg, which now contains the patient's DNA, begins to divide like a regular fertilized egg, it forms the early embryo, or the blastocyst ball of cells. Scientists would remove the stem cells and use them to grow any kind of tissue, nerve cell, or organ needed to treat the patient.

RELIGION AND POLITICS: SOLUTIONS?

In the United States in 2001, President George W. Bush put extreme limits on allowing federal money to finance research using embryonic stem cells if that research destroys the embryo, because the president believes that is equivalent to killing a person. The Catholic Church also teaches this belief, and other people feel the same way who believe that life begins at the moment of fertilization. Since fertilized eggs are usually destroyed in embryonic stem cell research, the president, as well as others of the same view, believes that embryonic stem cell research should not be permitted.

But new research may offer a solution to that dilemma. Scientists have been able to extract and grow embryonic stem cells without harming the embryo. They say they can pluck a single cell from a blastocyst of less than a dozen cells, leaving the embryo intact, and using that cell to create a line of stem cells. This new method is based on a procedure that is routinely used in fertility clinics to test lab-created embryos for genetic disorders.

"It's relatively simple and does not injure the embryo. In fact, it's been used for years to generate thousands of healthy babies," says Dr. Robert Lanza, medical director and research team head at Advanced Cell Technology in Worcester, Massachusetts.

"During this procedure, a single cell is normally removed from an eight-cell stage embryo and then sent off to the laboratory for genetic testing."

What Lanza is proposing is that they simply let that cell grow overnight, send one of the resulting cells off for testing, and then use the remaining cell or cells to create stem cells. "Thus, it will not interfere at all with the clinical outcome in any way. It will not affect the chances of the couple having a child. So there's absolutely no added risk for the embryo, but there's very much a positive benefit.

"These stem cells would also be of value to the siblings as well as to the scientific community, who of course could use these since they are immortal and would continue to grow indefinitely. So again, our main goal here would be to increase the number of stem cell lines that are available for federal funding and thus give this field a badly needed jump-start."

Would this new technique, which does not harm the embryo, change the mind of those against embryonic stem cell research? Lanza is not sure. "Embryonic stem cell research has been synonymous with the destruction of an embryo for so long, it's hard to imagine those individuals coming around, at least in the short term, but I think it really does remove the last rational objection to stem cell research."

The removal of cells for genetic testing in in vitro fertilization labs has been going on for many years. It's quite common. So why did

Lanza and his colleagues have success where others have not? Why didn't anyone think of this before?

"It turns out [that] to derive embryonic stem cells is actually pretty tricky," he points out, even for the most experienced researchers. "Those early stages of derivation—it's just very easy to lose the stem cells. And what we're talking about here is actually removing a single cell even at an earlier stage. You're talking about normally generating cells from a blastocyst—a clump of cells that's anywhere from sixty-four to two hundred cells—whereas the embryo we're removing this from is only eight cells."

The key to his success, Lanza believes, is putting the removed cell into a petri dish with other similar cells. "People in the past were just pulling it out of the embryo and throwing it in a plastic dish, and of course, the cell isn't happy and it dies." How do you make an embryonic stem cell happy? Give it companionship. "We know that embryos like to be together. If you put two or three embryos in a dish, for instance, they will grow much more strongly. The same with embryonic stem cells. If you only have a few, they tend to die. They're social creatures." So it's not surprising that a cell that's talking with its neighbors—that is, it's trying to decide how to become a full organism with all the organs and tissues—that if it's plucked out and put in a dish, it has been shut off from communicating, and it dies. "It can't talk to any neighboring cells, and it doesn't know what to do."

AMNIOTIC STEM CELLS

Just a few months after the Lanza announcement came word, in January 2007, of an even more intriguing stem cell advance: the isolation and growth of stem cells taken from amniotic fluid. "They appear to have the ability to change into many types of tissue in a similar way that embryonic stem cells do," says Dr. Anthony Atala, professor and chair in the Department of Urology and director of the Institute for Regenerative Medicine at Wake Forest University

Baptist Medical Center in Winston-Salem, North Carolina. For almost a decade Atala and his colleagues wondered: "'Could we find a population of true stem cells in the amniotic fluid and the placenta?' And the answer was indeed, there was a true stem-cell population that could be found both in the amniotic fluid as well as the placenta, which is the tissue, of course, that surrounds the baby and the fluid during development."

They are not embryonic stem cells, says Atala. "They're a class of [their] own. They're neither embryonic nor adult, but they have properties of both." For example, they are closer to human embryonic stem cells not only because they are so young "but [also because] like human embryonic stem cells, the cells are able to grow very rapidly. They double in number every thirty-six hours, which is one of the advantages, of course, of human embryonic stem cells. We also found markers consistent with human embryonic stem cells, but to our surprise, after we started growing these cells, we also found markers consistent with adult stem cells."

Atala has been able to prod these stem cells into doing what they do best. He's been able to direct them to become different tissue types: nerve, liver, muscle, cartilage, bone, and others. But only in the laboratory. "This is still very early in its development stage. It's going to take a while before we're able to bring these technologies to patients."

But because these cells come from people, would tissue made from these cells be rejected in recipients? "That's actually another nice advantage of these cells," points out Atala. "The fact is that you can preserve these cells after birth or during birth, if they have an amniocentesis. You could preserve these cells for self-use. So you can just freeze these cells now, and then use these cells if there's ever any problem with that particular patient."

But there is a second way to use these cells, and that is by creating a cell bank that would be available to other patients for transplant, the way a kidney or liver or heart is available now. You would find a donor "match" among the thousands and thousands of cells in the cell bank, says Atala. "This has all been worked in the transplantation field. You would need a bank of approximately one hundred thousand specimens to be able to provide ninety-nine percent of the U.S. population with a perfect genetic match for transplantation.

"And that would be a lot easier to do with this source of cells, because, as you know, there are approximately 4.5 million births per year in the U.S. So every birth could be a potential source of stem cells from the afterbirth, or what we call the placenta, and we could just retrieve those cells, bank those cells, and then have those available for other patients when the need arose."

The ability of these amniotic stem cells to transform into other tissues does not mean that embryonic stem cell research should be curtailed, says Atala.

"These are different from human embryonic stem cells; these are different from adult stem cells. And we don't know yet, long term, which cell would be best for what particular application. I think all research is important because we just don't know what cell is going to be best for what particular disease process."

Atala is continuing to refine his research, poking his stem cells to function at the same level as the normal cell. "They are actually secreting the things they're supposed to be secreting, and they are hooking up together with each other like they're supposed to do. We have a lot of work to do still. Even though it's exciting, of course, it's still in its early stage of development."

PEOPLE: A CLONING NO-NO

While there has been talk from rogue scientists about cloning people, and while there is no federal American law banning cloning, many scientists and physicians and organizations such as the American

Medical Association and the American Association for the Advancement of Science believe that human reproductive cloning is unwise, if not unethical. Most attempts to clone animals fail, with only one or two healthy offspring for every 100 experiments. Many cloned offspring who seem healthy die prematurely. And no one knows how cloning could affect mental and emotional development in humans.

AFTERWORD

GLOBAL WARMING GETS RELIGION

More and more, the environment and global warming are becoming less of a political and religious issue and more of a common sense one.

When Reverend Sally Bingham of Grace Cathedral in San Francisco told me that she was joining with congregations in almost two dozen other states and adopting the idea that stewardship of the earth *was* a moral issue, she was still a minority voice. Not many other denominations had joined her insistence that we are the earth's "caretakers" and it is our duty to "cut carbon emissions."

In the ensuing months, the positions of other clergy radically changed. In a major U-turn, dozens of Southern Baptist leaders, at their convention in March 2008, backed a resolution saying that too little had been done about climate change and that intense action was needed to mitigate it.

Their declaration is significant not only for its policy shift, but for the size of their flock. Normally a very conservative organization, the church leaders wrote that in the past they had been "too timid, failing to produce a unified moral voice." Looking forward, the leaders noted that they needed to become more involved. "Our cautious response to these issues in the face of mounting evidence may be seen by the world as uncaring, reckless and ill-informed."

In a document called "A Southern Baptist Declaration on the Environment and Climate Change," the ultra-conservative organization said that even though not all Christians accept the science behind global warming, the evidence was "substantial" and the potential for disaster too great to dismiss global warming as a liberal ruse. Action had to be taken immediately.

Southern Baptists comprise the second largest group of Christians in the United States, numbering 16 million, right behind the Roman Catholic Church.

Talk about a turnaround. Dare we think that, finally, the environment may be on its way to becoming a uniting rather than a dividing issue? Religion-wise, perhaps. Politically speaking, not quite there yet. Signs of cracks are beginning to show, however.

Many people are beginning to understand that their politics can be at odds with logic. Here's a good example. Earlier in the book when I spoke about energy policy, in chapter 11, I related the story of Ronald Reagan who ripped the solar panels installed by Jimmy Carter off the White House. It was a significant, symbolic policy shift. He was pointing the country in a new direction, away from conservation and energy independence. In my view, this act, and the cutting back of alternative energy research that followed, was the first fissure between science and politics. It signaled that science, in this case energy conservation and energy independence, would become a political issue as never before in America. From that day forward, conservatives would label anyone who was in favor of conservation and the environment as a "tree-hugging liberal." The label would stick for decades.

Fast forward to a recent incident as related to me by my wife, Miriam. She had just attended a meeting of school parents who had viewed Al Gore's film *An Inconvenient Truth*. A discussion about the message of the film ensued in which Miriam had related to the group my story about Carter and Reagan and the solar panels on the White House. I received a call from her, at the meeting, with just a bit of alarm in her voice. She had called to ask me if I was sure about the solar panels. I repeated what I had told her. As a reporter, I witnessed the events firsthand and was pretty sure of my recollection, even though it was more than thirty years ago. Okay, she said, that's what she needed to hear. She told me that a woman at the meeting came up to her and told her that she was a staunch Reagan conservative and idolized the past president. But she found it hard to believe that such an intelligent man would have done such a foolish thing as ripping the solar panels off the White House roof. Was my wife sure that such an event happened!? That led to the phone call to me and to my verification. I told my wife to go ahead and confirm what I had said, which she did. The woman stormed off and said she still didn't believe it and she would check it out for herself.

This event is a prime example of what I call the modern right brain–left brain disconnect. The emotional right side of the brain—the political, religious side—appears to be in conflict with the logical, objective left side of some people's brains. Their politics tell them not to believe in "tree hugger" ideas like global warming but their logic tells them the evidence is overwhelming. So they are conflicted and perhaps a bit confused. Which side should they obey?

Well, as it turns out, now they can listen to both sides; the environment and global warming may finally become a uniting issue. As the resolution by the Southern Baptist Convention frees up people to accept global warming from a religious point of view, so too many historical facts. My wife reports that about a month later, the woman at the meeting approached her and apologized. She had checked the facts and discovered that what I had said about her revered president was indeed true.

REPROGRAMMED STEM CELLS

A year can seem like a generation in the fast-moving world of stem cell research. Let's hope that by the time you read this, the research won't be "yesterday's news." It all began in late 2007 when Japanese and American researchers, Dr. Shinya Yamanaka at Kyoto University and Dr. James Thomson at the University of Wisconsin, Madison, reported a breakthrough: skin cells could be "reprogrammed" to revert back to their embryonic state.

If these reprogrammed cells share enough of the characteristics of embryonic stem cells, scientists will not need to harvest embryonic stem cells. It might solve a moral question some people have about creating embryos in the laboratory.

That's a big if. "They are not precisely like embryonic stem cells," says Dr. John Kessler, director of Northwestern University Stem Cell Institute. "We really have a lot of work to do before we understand exactly whether these cells will be safe to be used in human beings and, in fact, whether they can even be used in human beings."

Safe? What could go wrong? For one thing the methodology used to transform the cells carries risks. Scientists use viruses to insert specific genes into the skin cells. "Anytime you use a virus, you run the risk ultimately of that virus creating problems in the cells," points out Kessler.

And even if you could do away with using the viruses as carriers of the genes, the genes still pose a risk. "If you insert new genes into cells and they disrupt the normal pattern of DNA in the cell, you run the risk of converting that cell into something that would be a cancer cell. So we have a long way to go before we learn how to put these genes in, in a way that we don't disrupt the normal mechanisms of cell—of controlling cell division."

That work is moving forward. Other scientists have confirmed the technique, duplicating the results. But scientists warn that it is still too early to think about shelving conventional embryonic stem cell research since it has not yet been shown that the reprogrammed cells (known formally as induced pluripotent stem cells, or iPS cells) are effective replacements.

"Some people seem to believe that, now, we don't have to do any more embryonic stem cell work. In fact, we do. Those will be the gold standard by which we examine these new cells," says Kessler.

SOLAR THERMAL ENERGY: THE IMPENDING BOOM

You've seen solar thermal panels on roofs from California to Maine. Maybe you've even thought about getting some for your own home. But the heat from those hot water panels is a drop in the bucket, small potatoes, compared to the energy we could be harvesting from the sun. And now, after decades of experimenting in the wilderness —literally—solar thermal energy is ready for the big time. But we're not talking about panels on *your* rooftop. Rather, vast fields of panels reflecting, focusing, and collecting the sun's energy.

The idea is to centralize solar energy collection in one location and then feed it into your home, as electricity, just like the power company does now. It would be done using arrays of solar collectors grouped where the sun shines brightest and longest—the desert, and other large open spaces. The sun's heat would be used to boil water, creating steam and running a turbine to generate electricity. About a dozen solar thermal power plants are currently planned for California, Nevada, and Arizona.

One solar thermal company, Ausra, Inc., estimates that a square patch of land 100 miles on a side could supply enough electricity to power the entire country. That land need not be located all in one place, but could be spread out all over the country. Half the land area of the United States may be suitable for solar thermal power.

"The U.S. could nearly eliminate our dependence on coal, oil, and gas for electricity and transportation, drastically slashing global warming pollution without increasing costs for energy," said David Mills, chief scientific officer and founder at Ausra. "Our daily and annual energy needs closely match the energy production potential from solar thermal power plants with heat energy storage, and our models show solar thermal power will cost less than continuing to import oil."

The idea comes in a couple of flavors. One is a "tower" model. "It's a technology where very large fields of mirrors, the size of billboards, are used to focus the sun's energy on the top of a tower and then that heat is used to generate steam and run a power plant," says Fred Morse, senior advisor for U.S. operations of Abengoa Solar. Abengoa built the first solar tower in Spain and plans to build more of them around the world.

Another flavor is the "trough" idea. "If you want to visualize it," says Morse, "try to imagine a curved surface that's about 15 feet open at the aperture that stands about 15 feet high, and that's one-and-a-half football fields in length. So, it's a very long, mirrored surface. And down the center of that, where the sun's rays are focused, is a large glass pipe and it's evacuated so that, like a roach motel, all the sun's rays that get in don't get out.

"You can achieve very high temperatures—well over 700 degrees Fahrenheit—and that's hot enough to make steam to run a steam turbine."

So what happens at night? The trick is to store enough hot water in giant tanks—like super Thermos bottles—so that the plant can make electricity when the sun doesn't shine.

Abengoa is building such a plant, a 250-megawatt giant, in Arizona. It "will have three square miles of solar field and will have six hours of storage, which means when the sun goes down in Phoenix in August and the residents still have their air conditioners on at 10 or 11 at night, this solar plant will still be providing energy."

Another type does not use a trough but rather mounts special Fresnel mirrors above the ground. They track the sun, adjusting for the best angle as the sun moves across the sky. "That allows us to make the essential minimum structure very large," points out Mills. "Our absorber lines are about 1,300 feet long and about 100 feet wide. It's not something you put on your roof, but it is very efficient at collecting energy for very little expenditure on structural materials."

California's Pacific Gas and Electric Company has already signed up Ausra for such a solar utility, looking to purchase 177 megawatts of solar thermal power from a plant located in central California. The price per kilowatt should be very competitive with fossil fuels.

"We're starting to look at forward costs of natural gas generation of the order of 9 cents a kilowatt-hour in California according to the California government," says Mills. "And we believe this technology is only slightly above that, and is on a trajectory which is going to pass through that fairly shortly."

Plants on the order of 200 megawatts would not be out of the ordinary. It would take less than a dozen of these plants to replace the electricity produced by three nuclear power plants. This is a great advantage in many ways.

First, it takes years to site and build a nuclear power plant. Few people want them in their backyards. Yet the price of oil and natural gas continues to rise. Second, solar produces its peak power at the peak demand: in the good old summertime, when air conditioners are running full blast.

As wind farms sprout up around the world, solar utilities may work well to complement the times when the winds die down, an idea being investigated at the National Renewable Energy Laboratory (NREL) in Golden, Colorado. Mark Mehos, program manager, says that integrating the growing solar energy utility industry with the already blossoming wind farm utilities is something they are already studying. "When the wind does die off, the output from any particular plant also dies, and that can cause some problems on the grid. Solar has the advantages of having thermal storage. So, when the sun does go down late in the evening or if there is a cloud that comes over, our power doesn't go out—we can keep operating that particular plant."

From all accounts, it appears that solar power utilities are an idea whose time has come. In quiet development for the past 20 years, their time to shine appears to be now.

"We had a 15-year pause from the last plants that were built in the Mojave [decades ago]," says Morse. "We have to get the plants that have been signed up for—like ours, like Ausra's, like other's—built. And when we build them, we will learn how to build them faster, more efficiently. We will find ways to improve the technology. The research that NREL is doing will be brought into the market by these various companies. As we compete, the cost will come down and at the same time any carbon policy will raise the price of the competition, and I think you will see thousands of megawatts in the southwest being built annually in a few years." Look for next generation solar collectors in Europe, in Africa, in China, and maybe in a field near you.

✳ ACKNOWLEDGMENTS

It is my good fortune to know many smart and talented people whose work on my behalf made this book possible. Foremost are my stalwart producers on *Science Friday*—Charles Bergquist, Annette Heist, and Karin Vergoth—who spend the many hours preparing the interviews that take place on NPR's *Talk of the Nation: Science Friday* each week. It is those interviews that served as the backbone for this book.

My eyes were opened to the mythology of why airplanes fly by Norman Smith, but I could never quite nail down the physics until I read the works of Dr. David Anderson. He was good enough to send me a PowerPoint presentation for a lecture he gave at Fermilab on the physics of flight, as well as his terrifically illustrated book *Understanding Flight*, written with coauthor Scott Eberhardt. One look, and I had the solid grounding in science I needed to overturn that myth, which had been festering in my mind ever since I had met Smith many years ago.

For two key illustrations—the double-walled nanotube jacket cover and the rare photo of a jet slicing through the clouds—I'm indebted to Chris Ewels for his out-of-this-world carbon nanotube illustrations and to Paul Bowen for the permission to use his unique jet-in-the-clouds photo.

ACKNOWLEDGMENTS

This book owes its publication to my agent, Jonathan Lazear, who sat through many years of conversations with different publishers, finally successfully hooking up with Mathew Benjamin at HarperCollins.

INDEX

✳

Page numbers in *italics* indicate illustrations.

INDEX

Apache Web server, 321
aphasia, 28
Apollo space program, 174
Apple Computer, 258, 260–63
Apple I and Apple II computers, 260
Arab oil embargo (1970s), 102
archeology, 240
Argonne National Laboratory,
    123–24, *147*
Aristotle, 216–17
arterial disease, 17
arthritis, rheumatoid, 298, 333
astronomy, 57–58, 59, 61, 67–68, 75
Atala, Anthony, 335–37
atheists, 206, 214, 216, 219, 220–21
atomic force microscope, 155
atoms, 4, 81, 82, 279–80, 285, 287
Atwood, Margaret, 161
Auburn University, 112
Auden, W. H., 24
Ausra, 343, 346
authropic principle, 217
autoimmune disease, 294, 296, 297–98
automobiles, 103, 105–7
    hybrid, 113
    hydrogen fuel, 133, *147*
Awakenings (film), 23, 24, 28
axions, 69

bacteria, 162, 168–69, 318
Bahcall, Neta, 67–68
"Balkanzing effect," 273–74
bandwidth, 268–70, 271, 276
Banks Island Wind Farm, 144
barrier islands, 86, 88, 89
basal ganglia, 27
Bat Conservation International, 150
bats, 150
batteries, bacteria-powered, 169
Bea, Robert, 88–89, 91
beaches, 86, 88, 89
beauty, 291–92
beer bubbles, 313–15, *315*
behavioral cognitive group therapies,
    40
Behlendorf, Brian, 321
Belcher, Angela, 167–68
Benedek, László, 7–8
Benkler, Yochai, 321

Bennett, Charles L., 67
Berger, Hans, 8
Berkeley, Edmund C., 101
Bernoulli, Daniel, x
Bernoulli's principle, x–xiii, 304–5,
    306, 309
big bang, 54–55, 59, 66, 67, 69, 74, 79
Big Coal (Tidwell), 133
Bigelow, Robert, 178, 184
Bigfoot, 245, 246
Biggert, Judy, 122–24
binary computing, 280
Bingham, Sally, 93, 94–97, 339
biodiesel fuel, 105–6, 107, 113–15
biodiversity, oceanic, 192, 194,
    198–99
bioelectricity, 4–5, 11
biofuels. *See* biodiesel fuel; ethanol
biological warfare agents, 159
biology, 164, 286, 328
    open-source model and, 316–20
bionanotechnology, 158–59
BIOS initiative, 317
Bishop, Jerry, xii
bits, 280–81
black holes, 74, 284–85
blastocysts, *331*, 332–35
blogs, 271, 272
Boing Boing (blog), 271
Boltzmann, Ludwig, 279–80
Boltzmann equation, 280
Bongo, Omar, 248
brain, 3–29
    addiction and, 30–42
    adolescence and, 6–7
    aging and, 15, 17–21
    color perception and, 21–22
    consciousness and, 3, 253
    development of, 5–7
    genetics and, 11, 19
    imaging of, 7–12, 27, 36, 42, 46
    information processing by, 14, 16,
        17
    injuries to, 4, 8, 13, 25, 48
    memory and. *See* memory
    neuron regeneration and, 13
    reward system and, 33–37
    sex-based differences in, 13–14
    sleep's effects on, 5, 6, 7, 44–48

INDEX

INDEX

INDEX

INDEX

National Science Foundation, 12
National Security Agency, 282, 283, 286
National Sleep Foundation, 6
Native Americans, 245, 249
natural gas, 149
Natural History Museum (London), 245
Natural Resources Defense Council, 196
natural selection, 227
nature, God vs., 212–18
Nature Conservancy, 195–96
Nelson, Aaron P., 17–21
Netherlands, 87–91
network neutrality, 270–72, 273
neurons, 4–5, 6, 11, 34
    information processing by, 14, 16, 17
    regeneration of, 13
neuroscience, 3, 12
neurotransmitters, 33–34
Nevada, 119, 120–23, 135
New Orleans, 85, 86, 88–91
Newton, Isaac, xii, 22, 72, 306
New York, 103, 109–10
New York Times, 222
nicotine addiction. *See* addiction
nitrogen oxide, 133, 134
Nobel laureates, 7, 54, 59, 154, 156, 203, 206, 300
North Dakota, 145
North Korea, 127
Northwestern University, 342
Novalis, 23, 24
NRC. *See* Nuclear Regulatory Commission
NREL. *See* National Renewable Energy Laboratory
NSA. *See* National Security Agency
Nuclear Energy Institute, 122
nuclear fusion, *117*
nuclear power, 116–31, 140
Nuclear Regulatory Commission, 120–21, 124–25, 129
nuclear weapons, 123–27
nutrition, 19, 20, 21

Oak Ridge Associated Universities, 120
Oak Ridge National Laboratory, 105, 111, 112
obesity, 36
oceans, *87,* 191–200. *See also* sea levels
    Earle, Sylvia and, 191–96
Of Pandas and People (textbook), 226, 227, 228
oil, 102, 103, 105–6, 134, 248
Olduvai Gorge, 239, 240
Omidyar, Pierre, 271–72
open-source model, 316–21
Orentzle, Richard, 309
organ transplants, 330–31, 333
Orion Crew Exploration Vehicle, *173, 174,* 175
ORNL. *See* Oak Ridge National Laboratory
Oryx and Crake (Atwood), 161
overeating, 35, 36, 39
Ozawa (Japanese politician), 123

pancreas, 294–96
Parkinson's disease, 23, 24, 27, 330, 333
parthenogenesis, 327
particle accelerators, 64, 66, 69
particle physics, 59, 62, 66, 69–70
Pasteur, Louis, 241, 293
patents, 317, 318, 320
patient-specific cells, 255–56
PCs. *See* personal computers
pebble-bed reactors, 127–30
Pennsylvania. *See* Dover School Board case
Penrose, Roger, 52, 54, 55
personal computers, 257–64, 267, 280, 316
Peterson, Larry, 272, 273, 275
pets, cloning of, 254, 326, 328–29
PET scans, 9, 12
Pew Internet Project, 269
pharmaceutical companies, 11, 38–39, 42
phishing, 272
photonics, 158
photoreceptors, 200
photovoltaic generators, 145

INDEX